T0320734

KINKS AND DOMAIN WALLS
An Introduction to Classical and Quantum Solitons

Kinks and domain walls are the simplest kind of solitons and hence they are invaluable for testing various ideas and for learning about non-perturbative aspects of field theories. They are the subject of research in essentially every branch of physics, ranging from condensed matter to cosmology.

This book is a pedagogical introduction to kinks and domain walls and their principal classical and quantum properties. The book starts out by discussing classical solitons, building up from examples in elementary systems to more complicated settings. The quantum properties are introduced, together with discussion of the very fundamental role that solitons may play in particle physics. The formation of solitons in phase transitions, their dynamics, and their cosmological consequences are further discussed. The book closes with an explicit description of a few laboratory systems containing solitons.

Kinks and Domain Walls includes several state-of-the-art results (some previously unpublished), providing a handy reference. Each chapter closes with a list of open questions and research problems. This book will be of great interest to both graduate students and academic researchers in theoretical physics, particle physics and condensed matter physics.

This title, first published in 2007, has been reissued as an Open Access publication on Cambridge Core.

TANMAY VACHASPATI is a professor in the Physics Department at Case Western Reserve University. He was the Rosenbaum Fellow for the Topological Defects Programme at the Isaac Newton Institute, and is a Fellow of the American Physical Society. Professor Vachaspati co-edited *The Formation and Evolution of Cosmic Strings* with Professors Gary Gibbons and Stephen Hawking.

KINKS AND DOMAIN WALLS

An Introduction to Classical and Quantum Solitons

TANMAY VACHASPATI

Case Western Reserve University

CAMBRIDGE
UNIVERSITY PRESS

CAMBRIDGE
UNIVERSITY PRESS

Shaftesbury Road, Cambridge CB2 8EA, United Kingdom

One Liberty Plaza, 20th Floor, New York, NY 10006, USA

477 Williamstown Road, Port Melbourne, VIC 3207, Australia

314–321, 3rd Floor, Plot 3, Splendor Forum, Jasola District Centre, New Delhi – 110025, India

103 Penang Road, #05–06/07, Visioncrest Commercial, Singapore 238467

We share the University's mission to contribute to society through the pursuit of a department of the University of Cambridge.

We share the University's mission to contribute to society through the pursuit of education, learning and research at the highest international levels of excellence.

www.cambridge.org
Information on this title: www.cambridge.org/9781009290418

DOI: 10.1017/9781009290456

First published 2007
Reissued as OA 2022

A catalogue record for this publication is available from the British Library.

ISBN 978-1-009-29041-8 Hardback
ISBN 978-1-009-29042-5 Paperback

Contents

Preface

Solitons were first discovered by a Scottish engineer, J. Scott Russell, in 1834 while riding his horse by a water channel when a boat suddenly stopped. A hump of water rolled off the prow of the boat and moved rapidly down the channel for several miles, preserving its shape and speed. The observation was surprising because the hump did not rise and fall, or spread and die out, as ordinary water waves do.

In the 150 years or so since the discovery of Scott Russell, solitons have been discovered in numerous systems besides hydrodynamics. Probably the most important application of these is in the context of optics where they can propagate in optical fibers without distortion: they are being studied for high-data-rate (tera-bits) communication. Particle physicists have realized that solitons may also exist in their models of fundamental particles, and cosmologists have realized that such humps of energy may be propagating in the far reaches of outer space. There is even speculation that all the fundamental particles (electrons, quarks etc.) may be viewed as solitons owing to their quantum properties, leading to a "dual" description of fundamental matter.

In this book I describe the simplest kinds of solitons, called "kinks" in one spatial dimension and "domain walls" in three dimensions. These are also humps of energy as in Scott Russell's solitons. However, they also have a topological basis that is absent in hydrodynamical solitons. This leads to several differences e.g. water solitons cannot stand still and have to propagate with a certain velocity, while domain walls can propagate with any velocity. Another important point in this regard is that strict solitons, such as those encountered in hydrodynamics, preserve their identity after scattering. The kinks and domain walls discussed in this book do not necessarily have this property, and can dissipate their energy on collision, and even annihilate altogether.

Why focus on kinks and domain walls? Because they are known to exist in many laboratory systems and may exist in other exotic settings such as the early universe. They provide a simple setting for discussing non-linear and non-perturbative physics. They can give an insight into the dynamics of phase transitions. Lessons learned from the study of kinks and domain walls may also be applied to other more complicated topological defects. Domain walls are good pedagogy as one can introduce novel field theoretic, cosmological, and quantum issues without extraneous complexities that occur with their higher co-dimension defects (strings and monopoles).

The chapters of this book can be approximately categorized under four different headings. The first two chapters discuss solitons as classical solutions, the next three describe their microscopic classical and quantum properties, followed by another three chapters that discuss macroscopic properties and applications. The very last chapter discusses two real-world systems with kinks and, very briefly, Scott Russell's soliton. The book should be accessible to a theoretically inclined graduate student, and a large part of the book should also be accessible to an advanced undergraduate. At the end of every chapter, I have listed a few "open questions" to inspire the reader to take the subject further. Some of these questions are intentionally open-ended so as to promote greater exploration. Needless to say, there are no known answers to most of the open questions (that is why they are "open") and the solutions to some will be fit to print.

Every time I think about research in this area, I feel very fortunate for having unwittingly chosen it, for my journey on the "soliton train" has weaved through a vast landscape of physical phenomena, each with its own flavor, idiosyncrasies, and wonder. I hope that this book, as it starts out in classical solitons, then moves on to quantum effects, phase transitions, gravitation, and cosmology, and a bit of condensed matter physics, has captured some of that wonder for the reader.

This is not the first book on solitons and hopefully not the last one either. In this book I have presented a rather personal perspective of the subject, with some effort to completeness but focusing on topics that have intrigued me. Throughout, I have included some material that is not found in the published literature. Prominent among these is Section 4.5, where it is shown that the leading quantum correction to the kink mass is negative. The discussion of Section 6.5, with its emphasis on a bifurcation of correlation scales, also expresses a new viewpoint. I had particular difficulty deciding whether to include or omit discussion of domain walls in supersymmetric theories. On the one hand, many beautiful results can be derived for supersymmetric domain walls. On the other, the high degree of symmetry is certainly not realized (or is broken) in the real world. Also, non-supersymmetric domain walls are less constrained by symmetries and hence have richer possibilities. In the end, I decided not to include a discussion of supersymmetric walls, noting the excellent review by David Tong (see below). Some other must-read references are:

1. Rajaraman, R. (1982). *Solitons and Instantons.* Amsterdam: North-Holland.
2. Rebbi, C. and Soliani, G. (1984). *Solitons and Particles.* Singapore: World Scientific.
3. Coleman, S. (1985). *Aspects of Symmetry.* Cambridge: Cambridge University Press.
4. Vilenkin, A. and Shellard, E. P. S. (1994). *Cosmic Strings and Other Topological Defects.* Cambridge: Cambridge University Press.
5. Hindmarsh, M. B. and Kibble, T. W. B. (1995). Cosmic Strings. *Rep. Prog. Phys.*, **58**, 477–562.
6. Arodz, H., Dziarmaga, J. and Zurek, W. H., eds. (2003). *Patterns of Symmetry Breaking.* Dordrecht: Kluwer Academic Publishers.
7. Volovik, G. E. (2003). *The Universe in a Helium Droplet.* Oxford: Oxford University Press.
8. Manton, N. and Sutcliffe, P. (2004). *Topological Solitons.* Cambridge: Cambridge University Press.
9. Tong, D. (2005). *TASI Lectures on Solitons*, [hep-th/0509216].

I am grateful to a number of experts who, over the years, knowingly or unknowingly, have shaped this book. Foremost among these are my physicist father, Vachaspati, and Alex Vilenkin, my Ph.D. adviser. This book would not have been written without the support of other experts who have collaborated with me in researching many of the topics that are covered in this book. These include: Nuno Antunes, Harsh Mathur, Levon Pogosian, Dani Steer, and Grisha Volovik. Over the years, several of the sections in the book have been influenced by conversations with, and in some cases, owe their existence to, Sidney Coleman, Gary Gibbons, Tom Kibble, Hugh Osborn, and Paul Sutcliffe. The book would have many more errors, were it not for comments by Harsh Mathur, Ray Rivers, Dejan Stojkovic, Alex Vilenkin, and, especially, Dani Steer who painstakingly went over the bulk of the manuscript, making it much more readable and correct. My colleagues, Craig Copi and Pete Kernan, have provided invaluable computer support needed in the preparation of this book. I thank the editorial staff at C.U.P. for their patience and professionalism, and the Universities of Paris (VII and XI) and the Aspen Center for Physics for providing very hospitable and conducive working environments.

As I learned, writing a book takes a lot of sacrifices, and my admiration for my family, with their happy willingness to tolerate this effort, has increased many-fold. This book could not have been written without the unflinching support of my wife, Punam, and the total understanding of my children, Pranjal and Krithi.

1

Classical kinks

Kink solutions are special cases of "non-dissipative" solutions, for which the energy density at a given point does not vanish with time in the long time limit. On the contrary, a dissipative solution is one whose energy density at any given location tends to zero if we wait long enough [35],

$$\lim_{t\to\infty} \max_{\mathbf{x}}\{T_{00}(t, \mathbf{x})\} = 0, \quad \text{dissipative solution} \tag{1.1}$$

where $T_{00}(t, \mathbf{x})$ is the time-time component of the energy-momentum tensor, or the energy density, and is assumed to satisfy $T_{00} \geq 0$. Dissipationless solutions are special because they survive indefinitely in the system.

In this book we are interested in solutions that do not dissipate. In fact, for the most part, the solutions we discuss are static, though in a few cases we also discuss field configurations that dissipate. However, in these cases the dissipation is very slow and hence it is possible to treat the dissipation as a small perturbation. In addition to being dissipationless, kinks are also characterized by a topological charge. Just like electric charge, topological charge is conserved and this leads to important quantum properties.

In this chapter, we begin by studying kinks as classical solutions in certain field theories, and devise methods to find such solutions. The simplest field theories that have kink solutions are first described to gain intuition. These field theories are also realized in laboratory systems as we discuss in Chapter 9. The simple examples set the stage for the topological classification of kinks and similar objects in higher dimensions (Section 1.10), and are valuable signposts in our discussion of the more complicated systems of Chapter 2.

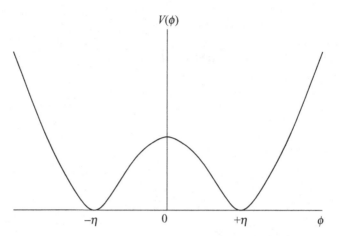

Figure 1.1 Shape of the $\lambda\phi^4$ potential.

1.1 Z_2 kink

The prototypical kink is the so-called "Z_2 kink." It is based on a field theory with a single real scalar field, ϕ, in $1 + 1$ dimensions. The action is

$$S = \int d^2x \left[\frac{1}{2}(\partial_\mu\phi)^2 - V(\phi) \right]$$

$$= \int d^2x \left[\frac{1}{2}(\partial_\mu\phi)^2 - \frac{\lambda}{4}(\phi^2 - \eta^2)^2 \right] \tag{1.2}$$

where $\mu = 0, 1$, and λ and η are parameters. The Lagrangian is invariant under the transformation $\phi \to -\phi$ and hence possesses a "reflectional" Z_2 symmetry. The potential for ϕ (see Fig. 1.1) is

$$V(\phi) = \frac{\lambda}{4}(\phi^2 - \eta^2)^2 = -\frac{m^2}{2}\phi^2 + \frac{\lambda}{4}\phi^4 + \frac{\lambda\eta^4}{4} \tag{1.3}$$

where $m^2 \equiv \lambda\eta^2$. The potential has two minima: $\phi = \pm\eta$, that are related by the reflectional symmetry. The "vacuum manifold," labeled by the classical field configurations with lowest energy, has two-fold degeneracy since $V(\phi) = V(-\phi)$.

The equations of motion can be derived from the action

$$\partial_t^2\phi - \partial_x^2\phi + \lambda(\phi^2 - \eta^2)\phi = 0 \tag{1.4}$$

where $\partial_t \equiv \partial/\partial t$ and similarly for ∂_x. A solution is $\phi(t, x) = +\eta$, and another is $\phi(t, x) = -\eta$. These have vanishing energy density and are called the "trivial vacua." The action describing excitations (sometimes called "mesons") about one of the trivial vacua can be derived by setting, for example, $\phi = \eta + \psi$, where ψ is the excitation field. Then

$$S = \int d^2x \left[\frac{1}{2}(\partial_\mu \psi)^2 - \frac{m_\psi^2}{2}\psi^2 - \sqrt{\frac{\lambda}{2}}m_\psi \psi^3 - \frac{\lambda}{4}\psi^4 \right] \quad (1.5)$$

where

$$m_\psi = \sqrt{2}m \quad (1.6)$$

is the mass of the meson.

Next consider the situation in which different parts of space are in different vacua. For example, $\phi(t, -\infty) = -\eta$ and $\phi(t, +\infty) = +\eta$. In this case, the function $\phi(t, x)$ has to go from $-\eta$ to $+\eta$ as x goes from $-\infty$ to $+\infty$. By continuity of the field there must be at least one point in space, x_0, such that $\phi(t, x_0) = 0$. Since $V(0) \neq 0$, there is potential energy in the vicinity of x_0, and the energy of this state is not zero. The solution of the classical equation of motion that interpolates between the different boundary conditions related by Z_2 transformations is called the "Z_2 kink."

We might wonder why the Z_2 kink cannot evolve into the trivial vacuum? For this to happen, the boundary condition at, say, $x = +\infty$ would have to change in a continuous way from $+\eta$ to $-\eta$. However, a small deviation of the field at infinity from one of the two vacua costs an infinite amount of potential energy. This is because as ϕ is changed, the field in an infinite region of space lies at a non-zero value of the potential (see Fig. 1.1). Hence, there is an infinite energy barrier to changing the boundary condition.[1]

A way to characterize the Z_2 kink is to notice the presence of a conserved current

$$j^\mu = \frac{1}{2\eta}\epsilon^{\mu\nu}\partial_\nu \phi \quad (1.7)$$

where $\mu, \nu = 0, 1$ and $\epsilon^{\mu\nu}$ is the antisymmetric symbol in two dimensions ($\epsilon^{01} = 1$). By the antisymmetry of $\epsilon^{\mu\nu}$, it is clear that j^μ is conserved: $\partial_\mu j^\mu = 0$. Hence

$$Q = \int dx j^0 = \frac{1}{2\eta}[\phi(x = +\infty) - \phi(x = -\infty)] \quad (1.8)$$

is a conserved charge in the model. For the trivial vacua $Q = 0$, and for the kink configuration described above $Q = 1$. So the kink configuration cannot relax into the vacuum – it is in a sector that carries a different value of the conserved "topological charge."

To obtain the field configuration with boundary conditions $\phi(\pm\infty) = \pm\eta$, we solve the equations of motion in Eq. (1.4). We set time derivatives to vanish since

---

[1] In Chapter 2 we will come across an example where the vacuum manifold is a continuum and correspondingly there is a continuum of boundary conditions that can be chosen as opposed to the discrete choice in the Z_2 case. This will lead to some new considerations.

we are looking for static solutions. Then, the kink solution is

$$\phi_k(x) = \eta \, \tanh\left(\sqrt{\frac{\lambda}{2}}\eta x\right) \qquad (1.9)$$

In fact, one can Lorentz boost this solution to get

$$\phi_k(t, x) = \eta \, \tanh\left(\sqrt{\frac{\lambda}{2}}\eta X\right) \qquad (1.10)$$

where

$$X \equiv \frac{x - vt}{\sqrt{1 - v^2}} \qquad (1.11)$$

(Recall that we are working in units in which the speed of light is unity i.e. $c = 1$.)
The solution in Eq. (1.10) represents a kink moving at velocity v.

Another class of solutions is obtained by translating the solution in Eq. (1.9)

$$\phi_k(x; a) = \eta \, \tanh\left(\sqrt{\frac{\lambda}{2}}\eta(x - a)\right) \qquad (1.12)$$

It is easily checked that translations do not change the energy of the kink. This is
often stated as saying that the kink has a zero energy fluctuation mode (or simply
a "zero mode"). To explain this statement, we need to consider small fluctuations
of the field about the kink solution, similar to Eq. (1.5). We now have

$$\phi = \phi_k(x) + \psi(t, x) \qquad (1.13)$$

where ϕ_k denotes the kink solution. The fluctuation field, ψ, obeys the linearized
equation

$$\partial_t^2 \psi - \partial_x^2 \psi + \lambda\left(3\phi_k^2 - \eta^2\right)\psi = 0 \qquad (1.14)$$

To find the fluctuation eigenmodes we set

$$\psi = e^{-i\omega t} f(x) \qquad (1.15)$$

where $f(x)$ obeys

$$-\partial_x^2 f + \lambda\left(3\phi_k^2 - \eta^2\right) = \omega^2 f \qquad (1.16)$$

We will discuss all the solutions to this equation in Chapter 4. Here we focus on the
translation mode. Since translations cost zero energy, there has to be an eigenmode
with $\omega = 0$. This can be obtained by directly solving Eq. (1.16) or by noting that
for small a, the solution in Eq. (1.12) can be Taylor expanded as

$$\phi_k(x; a) = \phi_k(x; a = 0) + a\frac{d\phi_k}{dx}\bigg|_{a=0} \qquad (1.17)$$

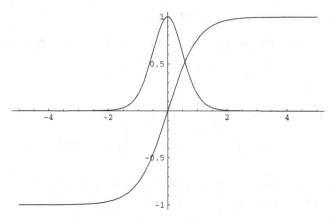

Figure 1.2   The curve ranging from -1 to $+1$ as x goes from $-\infty$ to $+\infty$ shows the Z_2 kink profile for $\lambda = 2$ and $\eta = 1$. The energy density of the kink has also been plotted on the same graph for convenience, and to show that all the energy is localized in the narrow region where the field has a gradient.

Comparing Eqs. (1.17) and (1.13), the zero mode solution is

$$f_0(x) = \left.\frac{d\phi_k}{dx}\right|_{a=0} = \eta^2\sqrt{\frac{\lambda}{2}}\,\mathrm{sech}^2\left(\sqrt{\frac{\lambda}{2}}\eta x\right) \tag{1.18}$$

The solution in Eq. (1.9) can be used to calculate the energy density of the kink

$$\begin{aligned}\mathcal{E} &= \frac{1}{2}(\partial_t\phi_k)^2 + \frac{1}{2}(\partial_x\phi_k)^2 + V(\phi_k)\\ &= 0 + V(\phi_k) + V(\phi_k)\\ &= \frac{\lambda\eta^4}{2}\mathrm{sech}^4\left(\sqrt{\frac{\lambda}{2}}\eta x\right)\end{aligned} \tag{1.19}$$

where the second line is written to explicitly show that $(\partial_x\phi)^2 = 2V(\phi)$. The kink profile and the energy density are shown in Fig. 1.2. The total energy is

$$E = \int dx\,\mathcal{E} = \frac{2\sqrt{2}}{3}\frac{m^3}{\lambda} \tag{1.20}$$

As is apparent from the solution and also the energy density profile, the half-width of the kink is,

$$w = \sqrt{\frac{2}{\lambda}\frac{1}{\eta}} = \frac{\sqrt{2}}{m} = \frac{2}{m_\psi} \tag{1.21}$$

On the $x > 0$ side of the kink we have $\phi \sim +\eta$ while on the $x < 0$ side we have $\phi \sim -\eta$. At the center of the kink, $\phi = 0$, and hence the Z_2 symmetry is restored in the core of the kink. Therefore the interior of the kink is a relic of the symmetric phase of the system.

1.2 Rescaling

It is convenient to rescale variables in the action in Eq. (1.2) as follows

$$\Phi = \frac{\phi}{\eta}, \qquad y^\mu = \sqrt{\lambda}\,\eta x^\mu \tag{1.22}$$

Then the rescaled action is

$$S = \eta^2 \int d^2y \left[\frac{1}{2}(\partial_\mu \Phi)^2 - \frac{1}{4}(\Phi^2 - 1)^2 \right] \tag{1.23}$$

where derivatives are now with respect to y^μ. The overall multiplicative factor, η^2, does not enter the classical equations of motion. Hence the classical $\lambda\phi^4$ action is free of parameters.[2]

1.3 Derrick's argument

In the context of rescaling, we now give Derrick's result that there can be no static, finite energy solutions in scalar field theories in more than one spatial dimension [45]. Consider the general action in n spatial dimensions

$$S = \int d^{n+1}x \left[\frac{1}{2}\sum_a (\partial_\mu \phi^a)^2 - V(\phi^a) \right] \tag{1.24}$$

where the potential is assumed to satisfy $V(\phi^a) \geq 0$. The index on ϕ^a means that the model can contain an arbitrary number of scalar fields. Let a purported static, finite energy solution to the equations of motion be $\phi_0^a(x^\mu)$ and consider the rescaled field configuration

$$\Phi_0^a(x^\mu) = \phi_0^a(\alpha x^\mu) \tag{1.25}$$

where $\alpha \geq 0$ is the rescaling parameter. Then the energy of the rescaled field configuration is

$$E[\Phi_0^a] = \int d^n x \left[\frac{1}{2}(\nabla \Phi_0^a)^2 + V(\Phi_0^a) \right] \tag{1.26}$$

where the sum over a is implicit and ∇ denotes the spatial gradient. Now define $y^\mu = \alpha x^\mu$ and this gives

$$E[\Phi_0^a] = \int d^n y \left[\frac{\alpha^{-n+2}}{2}(\nabla \phi_0^a(y))^2 + \alpha^{-n}\, V(\phi_0^a(y)) \right] \tag{1.27}$$

---

[2] In quantum theory, however, the value of the action enters the path integral evaluation of the transition amplitudes and this will depend on η^2. So the properties of the quantized kink also depend on the value of η^2 (see Chapter 4).

Since the kinetic terms are non-negative, we find that with $n \geq 2$ and $\alpha > 1$ this gives

$$E[\Phi_0^a] < E[\phi_0^a] \tag{1.28}$$

and hence ϕ_0^a cannot be an extremum of the energy. Only if $n = 1$ can ϕ_0^a be a static, finite energy solution.

In more than one spatial dimension, Derrick's argument allows for static solutions of infinite energy. The next section describes one such static solution in three spatial dimensions.

1.4 Domain walls

When kink solutions are placed in more than one spatial dimension, they become extended planar structures called "domain walls." The field configuration for a Z_2 domain wall in the yz-plane in three spatial dimensions is

$$\phi(t, x, y, z) = \eta \, \tanh\left(\sqrt{\frac{\lambda}{2}}\eta x\right) \tag{1.29}$$

The energy density of the wall is concentrated over all the yz-plane and is given by Eq. (1.19). The new aspects of domain walls are that they can be curved and deformations can propagate along them. These will be discussed in detail in Chapter 7.

Another feature of the planar domain wall is that it is invariant under boosts in the plane parallel to the wall. This is simply because the solution is independent of t, y and z and any transformations of these coordinates do not affect the solution.

1.5 Bogomolnyi method for Z_2 kink

Rather than directly solve the equations of motion, as was done in Section 1.1, we can also obtain the kink solution by the clever method discovered by Bogomolnyi [20]. The method is to obtain a first-order differential equation by manipulating the energy functional into a "whole square" form

$$\begin{aligned}
E &= \int dx \left[\frac{1}{2}(\partial_t\phi)^2 + \frac{1}{2}(\partial_x\phi)^2 + V(\phi)\right] \\
&= \int dx \left[\frac{1}{2}(\partial_t\phi)^2 + \frac{1}{2}\left(\partial_x\phi \mp \sqrt{2V(\phi)}\right)^2 \pm \sqrt{2V(\phi)}\partial_x\phi\right] \\
&= \int dx \left[\frac{1}{2}(\partial_t\phi)^2 + \frac{1}{2}\left(\partial_x\phi \mp \sqrt{2V(\phi)}\right)^2\right] \pm \int_{\phi(-\infty)}^{\phi(+\infty)} d\phi' \sqrt{2V(\phi')}
\end{aligned}$$

Then, for fixed values of ϕ at $\pm\infty$, the energy is minimized if

$$\partial_t \phi = 0 \tag{1.30}$$

and

$$\partial_x \phi \mp \sqrt{2V(\phi)} = 0. \tag{1.31}$$

Further, the minimum value of the energy is

$$E_{\min} = \pm \int_{\phi(-\infty)}^{\phi(+\infty)} d\phi' \sqrt{2V(\phi')}. \tag{1.32}$$

The energy can only be minimized provided a solution to Eq. (1.31) exists with the correct boundary conditions. This relates the choice of sign in Eq. (1.31) to the boundary conditions and to the sign in Eq. (1.32). In our case, for the Z_2 kink boundary conditions ($\phi(+\infty) > \phi(-\infty)$), we take the $-$ sign in Eq. (1.31). Inserting

$$\sqrt{V(\phi)} = \sqrt{\frac{\lambda}{4}(\eta^2 - \phi^2)} \tag{1.33}$$

in Eq. (1.31) we get the kink solution in Eq. (1.9).

The energy of the kink follows from Eq. (1.32)

$$E = \frac{2\sqrt{2}}{3}\sqrt{\lambda}\eta^3 = \frac{2\sqrt{2}}{3}\frac{m^3}{\lambda} \tag{1.34}$$

where $m = \sqrt{\lambda}\eta$ is the mass scale in the model (see Eq. (1.3)).

1.6 Z_2 antikink

In an identical manner, we can construct antikink solutions that have $Q = -1$. The boundary conditions necessary to get $Q = -1$ are $\phi(\pm\infty) = \mp\eta$ (see Eq. (1.8)). In the Bogomolnyi method, antikinks are obtained by taking the opposite choice of signs to the ones in the previous section

$$E = \int dx \left[\frac{1}{2}(\partial_t \phi)^2 + \frac{1}{2}(\partial_x \phi + \sqrt{2V(\phi)})^2 - \sqrt{2V(\phi)}\partial_x \phi \right] \tag{1.35}$$

This leads to the antikink solution

$$\bar{\phi}_k = -\eta \tanh\left(\sqrt{\frac{\lambda}{2}}\eta x\right) \tag{1.36}$$

1.7 Many kinks

The kink solution is well-localized and so it should be possible to write down field configurations with many kinks. However, a peculiarity of the Z_2 kink system is that a kink must necessarily be followed by an antikink since the asymptotic fields are restricted to lie in the vacuum: $\phi = \pm\eta$. It is not possible to have neighboring Z_2 kinks or a system with topological charge $|Q| > 1$.

There is a simple scheme, called the "product ansatz," to write down approximate multi-kink field configurations, i.e. alternating kinks and antikinks. Suppose we have kinks at locations $x = k_i$ and antikinks at $x = l_j$, where i, j label the various kinks and antikinks. The locations are assumed to be consistent with the requirement that kinks and antikinks alternate: $\ldots l_i < k_i < l_{i+1} \ldots$ Then an approximate field configuration that describes N kinks and N' antikinks is given by the product of the solutions for the individual objects with a normalization factor

$$\phi(x) = \frac{1}{\eta^{N+N'-1}} \prod_{i=1}^{N} \phi_k(x - k_i) \prod_{j=1}^{N'} [-\phi_k(x - l_j)] \tag{1.37}$$

where ϕ_k is the kink solution. Note that $|N - N'| \leq 1$ since kinks and antikinks must alternate.

The product ansatz is a good approximation as long as the kinks are separated by distances that are much larger than their widths. In that case, in the vicinity of a particular kink, say at $x = k_i$, only the factor $\phi(x - k_i)$ is non-trivial. All the other factors in Eq. (1.37) multiply together to give $+1$.

Another scheme to write down approximate multi-kink solutions is "additive" [109]. If ϕ_i denotes the ith kink (or antikink) in a sequence of N kinks and antikinks, we have

$$\phi(x) = \sum_{i=1}^{N} \phi_i \pm (N_2 - 1)\eta, \quad N_2 = N \pmod 2 \tag{1.38}$$

where the sign is $+$ if the leftmost object is a kink and $-$ if it is an antikink.

Neither the product or the additive ansatz yields a multi-kink solution to the equations of motion. Instead they give field configurations that resemble several widely spaced kinks that have been patched together in a smooth way. If the multi-kink configuration given by either of the ansätze is evolved using the equation of motion, the kinks will start moving due to forces exerted by the other kinks. In the next section we discuss the inter-kink forces.

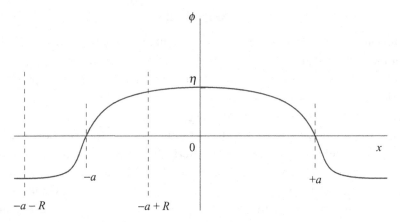

Figure 1.3   A widely separated kink-antikink.

1.8 Inter-kink force

Consider a kink at $x = -a$ and an antikink at $x = +a$ where the separation $2a$ is much larger than the kink width (see Fig. 1.3). We would like to evaluate the force on the kink owing to the antikink [109].

The energy-momentum tensor for the action Eq. (1.2) with a general potential $V(\phi)$ is

$$T_{\mu\nu} = \partial_\mu\phi\partial_\nu\phi - g_{\mu\nu}\left\{\frac{1}{2}(\partial_\alpha\phi)^2 - V(\phi)\right\} \tag{1.39}$$

where $g_{\mu\nu}$ is the metric tensor that we take to be the flat metric, that is, $g_{\mu\nu} = \text{diag}(1, -1)$. The force exerted on a kink is given by Newton's second law, by the rate of change of its momentum. The momentum of a kink can be found by integrating the kink's momentum density, $T^{0i} = -T_{0i}$, in a large region around the kink. If the kink is located at $x = -a$, let us choose to look at the momentum, P, of the field in the region $(-a - R, -a + R)$

$$P = -\int_{-a-R}^{-a+R} \mathrm{d}x\, \partial_t\phi\partial_x\phi \tag{1.40}$$

After using the field equation of motion (for a general potential) and on performing the integration, the force on the field in this region is

$$F = \frac{\mathrm{d}P}{\mathrm{d}t} = \left[-\frac{1}{2}(\partial_t\phi)^2 - \frac{1}{2}(\partial_x\phi)^2 + V(\phi)\right]_{-a-R}^{-a+R} \tag{1.41}$$

To proceed further we need to know the field ϕ in the interval $(-a - R, -a + R)$. This may be obtained using the additive ansatz given in Eq. (1.38) which we take as an initial condition

$$\phi(t = 0, x) = \phi_k(x) + \bar{\phi}_k(x) - \phi_k(\infty) \tag{1.42}$$

In addition, we assume that the kinks are initially at rest

$$\partial_t \phi \Big|_{t=0} = 0 \tag{1.43}$$

The expression for the force is further simplified by using the Bogomolnyi equation (Eq. (1.31)) which is satisfied by both ϕ_k and $\bar{\phi}_k$

$$(\partial_x \phi)^2 = 2V(\phi) \tag{1.44}$$

This gives

$$F = \left[-\partial_x \phi_k \partial_x \bar{\phi}_k + V(\phi_k + \bar{\phi}_k - \phi_k(\infty)) - V(\phi_k) - V(\bar{\phi}_k) \right]_{-a-R}^{-a+R} \tag{1.45}$$

The terms involving the potential can be expanded since the field is nearly in the vacuum at $x = -a \pm R$. Let us define

$$\phi_k^{\pm} = \phi_k(-a \pm R), \quad \bar{\phi}_k^{\pm} = \bar{\phi}_k(-a \pm R)$$
$$\Delta\phi_k^{\pm} = \phi_k(-a \pm R) - \phi_k(\pm\infty)$$
$$\Delta\bar{\phi}_k^{\pm} = \bar{\phi}_k(-a \pm R) - \bar{\phi}_k(-\infty) \tag{1.46}$$

(Note that the argument in the very last term is $-\infty$, independent of the signs in the other terms. This is because both $x = -a \pm R$ lie to the left of the antikink.) Also define

$$m_\psi^2 \equiv V''(\phi_k(\infty)) = V''(\bar{\phi}_k(\infty)) \tag{1.47}$$

Then the force is

$$F = -(\partial_x \phi_k^+ \partial_x \bar{\phi}_k^+ - \partial_x \phi_k^- \partial_x \bar{\phi}_k^-) + m_\psi^2(\Delta\phi_k^+ \Delta\bar{\phi}_k^+ - \Delta\phi_k^- \Delta\bar{\phi}_k^-) \tag{1.48}$$

Let us illustrate this formula for the Z_2 kink, where

$$\phi_k(x) = \eta \tanh(\sigma(x + a))$$
$$\bar{\phi}_k(x) = -\eta \tanh(\sigma(x - a)) \tag{1.49}$$

with $\sigma = \sqrt{\lambda/2}\, \eta$. Inserting these expressions in Eq. (1.48) and retaining only the leading order behavior gives

$$F = \frac{4m_\psi^4}{\lambda} e^{-m_\psi l} \tag{1.50}$$

where $l \equiv 2a$ is the kink separation. The force is attractive since it is acting on the kink at $x = -a$ and points toward the antikink at $x = +a$.

The result for the force could have been guessed from other considerations. The kinks are interacting by the exchange of massive scalars of mass m_ψ. As described in many quantum field theory texts [119] the force mediated by scalar interactions is the Yukawa force which goes like $\exp(-m_\psi l)$. The dimensionful prefactor of the force can be deduced on dimensional grounds while the numerical coefficient requires more detailed analysis.

1.9 Sine-Gordon kink

The sine-Gordon model is a scalar field theory in $1 + 1$ space-time dimensions given by the Lagrangian

$$L = \frac{1}{2}(\partial_\mu \phi)^2 - \frac{\alpha}{\beta^2}(1 - \cos(\beta\phi)) \tag{1.51}$$

The model is invariant under $\phi \rightarrow \phi + 2\pi n$ where n is any integer and thus possesses Z symmetry. The vacua are given by $\phi = 2\pi n/\beta$ and are labeled by the integer n.

As in the Z_2 case, the classical kink solutions can be found directly from the second-order equations of motion or by using the Bogomolnyi method (see Section 1.5). The kinks are solutions that interpolate between neighboring vacua. The unit charge kink solution is

$$\phi_k = \frac{4}{\beta} \tan^{-1}\left(e^{\sqrt{\alpha}x}\right) + \phi(-\infty) \tag{1.52}$$

where the inverse tangent is taken to lie in the interval $(-\pi/2, +\pi/2)$. The antikink with $\phi(-\infty) = 2\pi/\beta$ and $\phi(+\infty) = 0$ is obtained from Eq. (1.52) by replacing x by $-x$.

$$\bar{\phi}_k = \frac{4}{\beta} \tan^{-1}\left(e^{-\sqrt{\alpha}x}\right) + \phi(+\infty) \tag{1.53}$$

The width of the kink follows directly from these solutions and is $\sim 1/\sqrt{\alpha}$.

The energy of the kink also follows from Bogomolnyi's method (Eq. (1.32))

$$E_{sG} = \frac{8\sqrt{\alpha}}{\beta^2} \tag{1.54}$$

Defining $m_\psi = \sqrt{\alpha}$ – the mass of excitations of the true vacuum – and $\sqrt{\lambda} = \sqrt{\alpha}\beta$ we get

$$E_{sG} = 8\frac{m_\psi^3}{\lambda} \tag{1.55}$$

While the Z_2 and sine-Gordon kinks are similar as classical solutions, there are some notable differences. For example, it is possible to have consecutive sine-Gordon kinks whereas in the Z_2 case, kinks can only neighbor antikinks. In addition, the sine-Gordon system allows non-dissipative classical bound states of kink and antikink – the so-called "breather" solutions – while no such solutions are known in the Z_2 case (though see Section 3.1). The sine-Gordon kink is also much more amenable to a quantum analysis as we discuss in Chapter 4.

We can use the additive ansatz described in Section 1.7 to construct field configurations for many kinks. Specializing to a kink-kink pair ($\phi(-\infty) = 0$ to $\phi(+\infty) = 4\pi/\beta$) and a kink-antikink pair ($\phi(-\infty) = 0$ and back to $\phi(+\infty) = 0$), we have

$$\phi_{kk}(t, x) = \frac{4}{\beta}\left[\tan^{-1}\left(e^{\sqrt{\alpha}(x-a)}\right) + \tan^{-1}\left(e^{\sqrt{\alpha}(x-b)}\right)\right] \tag{1.56}$$

$$\phi_{k\bar{k}}(t, x) = \frac{4}{\beta}\left[\tan^{-1}\left(e^{\sqrt{\alpha}(x-a)}\right) + \tan^{-1}\left(e^{-\sqrt{\alpha}(x-b)}\right)\right] - \frac{2\pi}{\beta} \tag{1.57}$$

with $b > a$.

The additive ansatz described above gives approximate solutions to the equations of motion for widely separated kinks ($b - a >> 1/\sqrt{\alpha}$). A one-parameter family of exact, non-dissipative, breather solutions composed of a kink and an antikink is

$$\phi_b(t, x) = \frac{4}{\beta}\tan^{-1}\left[\frac{\eta\sin(\omega t)}{\cosh(\eta\omega x)}\right] \tag{1.58}$$

where $\eta = \sqrt{\alpha - \omega^2}/\omega$ and the $\tan^{-1}$ function is taken to lie in the range $(-\pi/2, +\pi/2)$. The frequency of oscillation, ω, is the parameter that labels the different breathers of the one-parameter family.

To see the breather as a bound state of a kink and an antikink, note that $\phi(t, \pm\infty) = 0$. Also, if $\eta \gg 1$, then $\phi(t, 0) \approx 2\pi/\beta$ during the time when $\eta\sin(\omega t) \gg 1$. Hence the breather splits up into a kink and an antikink for part of the oscillation period. For the remainder of the oscillation period, the kink and an antikink overlap and a clear separation cannot be made.

The constant energy of the breather is evaluated by substituting the solution at $t = 0$ (for convenience) in the sine-Gordon Hamiltonian with the result

$$E_b = \frac{16}{\beta^2}\sqrt{\alpha - \omega^2} = 2E_{sG}\sqrt{1 - \frac{\omega^2}{\alpha}} \tag{1.59}$$

As expected, when $\omega \to 0$, the breather energy is twice the kink energy.

As in Section 1.8 we can find the force on a kink owing to an antikink: from Eq. (1.48) the leading order behavior of the force is

$$F = \frac{20m_\psi^2}{\beta^2}e^{-m_\psi l} \tag{1.60}$$

where l is the kink separation.

1.10 Topology: π_0

The kinks in the Z_2 and sine-Gordon models can be viewed as arising purely
for topological reasons, as we now explain. A very important advantage of the
topological viewpoint is that it is generalizable to a wide variety of models and
can be used to classify a large set of solutions. When applied to field theories in
higher spatial dimensions, topological considerations are convenient in order to
demonstrate the existence of solutions such as strings and monopoles.

Consider a field theory for a set of fields denoted by Φ that is invariant under
transformations belonging to a group G. This means that the Hamiltonian of the
theory is invariant under G:

$$\mathcal{H}[\Phi] = \mathcal{H}[\Phi^g] \tag{1.61}$$

where $g \in G$ and Φ^g represents Φ after it has been transformed by the action of
g. The group G is a symmetry of the system, if Eq. (1.61) holds for every $g \in G$
and for every possible Φ. Now, let the Hamiltonian be minimized when $\Phi = \Phi_0$.
Then, from Eq. (1.61), it is also minimized with $\Phi = \Phi_0^g$ for any $g \in G$, and the
manifold of lowest energy states – "vacuum manifold" – is labeled by the set of
field configurations Φ_0^g. However, there will exist a subgroup (sometimes trivial),
H of G, whose elements do not move Φ_0:

$$\Phi_0^h = \Phi_0 \tag{1.62}$$

Hence, a group element $gh \in G$ acting on Φ_0 has the same result as g acting on
Φ_0 (because h acts first and does not move Φ_0). So, while the configuration Φ_0^g
has the same energy as Φ_0 for any g, not all the Φ_0^gs are distinct from each other.
The distinct Φ_0^gs are labeled by the set of elements $\{gh : h \in H\} \equiv gH$. The set
of elements $\{gH : g \in G\}$ are said to form a "coset space" and the set is denoted
by G/H; each element of the space is a coset (more precisely a "left coset" since
g multiplies H from the left). Therefore the vacuum manifold is isomorphic to the
coset space G/H.

We have so far connected the symmetries of the model to the vacuum manifold.
Now we discuss the tools for describing the topology of the vacuum manifold. This
will lead to a description of the topology of the vacuum manifold directly in terms
of the symmetries of the model.

The topology of a manifold, M, is classified by the homotopy groups, $\pi_n(M; x_0)$,
$n = 0, 1, 2,\ldots$ The idea is to consider maps from n-spheres to M, with the
image of an n-sphere in M containing one common base point, x_0 (see Fig. 1.4).
If two maps can be continuously deformed into each other, they are considered
to be topologically equivalent. In this way, the set of maps is divided into equivalence classes of maps, where each equivalence class contains the set of maps that

are continuously deformable into each other. The elements of $\pi_n(M;x_0)$ are the equivalence classes of maps from S^n to M with fixed base point. It is also possible to define (except for $n = 0$ as explained below) a suitable "product" of two maps: essentially the product of maps f and g (denoted by $g \cdot f$) and is defined to be "f composed with g" or "f followed by g." Then it is easily verified that the product is closed, associative, an identity map exists, and every map has an inverse. In mathematical language, $\forall\, f,\, g,\, h \in G$,

$$f \cdot g \in G$$
$$f \cdot (g \cdot h) = (f \cdot g) \cdot h$$
$$\exists\, e \in G \text{ such that } f \cdot e = e \cdot f = f$$
$$\exists\, f^{-1} \in G \text{ such that } f \cdot f^{-1} = f^{-1} \cdot f = e \tag{1.63}$$

Thus all the group properties are satisfied and $\pi_n(M;x_0)$ is a group.

Two homotopy groups with different base points, say $\pi_n(M;x_0)$ and $\pi_n(M;x_0')$, can be shown to be isomorphic and hence the reference to the base point is often dropped and the homotopy group simply written as $\pi_n(M)$. Mathematicians have calculated the homotopy groups for a wide variety of manifolds and this makes it very convenient to determine if a given symmetry breaking leads to a topologically non-trivial vacuum manifold [145, 3, 171].

In the case of kinks or domain walls, the field defines a mapping from the points $x = \pm\infty$ to the vacuum manifold. Hence the relevant homotopy "group" is $\pi_0(M;x_0)$, which contains maps from S^0 (a point) to M. Since the base point is fixed, the image of either of the two possible S^0s ($x = \pm\infty$) has to be x_0, and $\pi_0(M;x_0)$ is trivial. Even if we do not impose the restriction that the maps should have a fixed base point, it is not possible to define a suitable composition of maps. Therefore π_0 does not have the right group structure and should merely be considered as a set of maps from S^0 to the vacuum manifold. The exception occurs if $M = G/H$ is itself a group, which occurs when H is a normal subgroup of G, because then $\pi_0(M)$ can inherit the group structure of M. In this case, the product of two maps from S^0 to M can be defined to be the map from S^0 to the product of the two image points in M. Generally, however, $\pi_0(M)$ should simply be thought of as a set of maps from S^0 to the various disconnected pieces of M.

To connect the elements of the homotopy groups to topological field configurations assume that the field, Φ, is in the vacuum manifold on S_∞^n. Therefore, $\Phi_\infty \equiv \Phi(\mathbf{x} \in S_\infty^n)$ defines a map from S^n to the vacuum manifold and this map can be topologically non-trivial if $\pi_n(M)$ is non-trivial. We want to show that if the map Φ_∞ is topologically non-trivial, Φ cannot be in the vacuum manifold at all points in the interior of S_∞^n. Consider what happens as the radius of S_∞^n is continuously decreased. If the field remains on the vacuum manifold, continuity implies that the

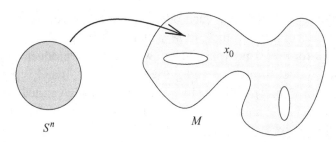

Figure 1.4   The nth homotopy group consists of maps from the n-dimensional sphere, S^n, to the vacuum manifold, M, such that the image of any map contains one common base point $x_0 \in M$. If two maps can be continuously deformed into each other, they are identified, and correspond to the same element of π_n. If two maps cannot be continuously deformed into each other, then they correspond to distinct elements of π_n. For example, this can happen if one of the maps encloses a "hole" in M, while the other encloses the hole a different number of times.

map Φ_R from a sphere of radius R to M must also be non-trivial. Then as $R \to 0$, the map would still be non-trivial, implying that the field is multivalued at the origin since the field must continue to map S_R^n non-trivially as $R \to 0$. However, a field (by definition) cannot be multivalued. The only way out is if the field does not lie on the vacuum manifold everywhere. Therefore non-trivial topology at infinity implies that the energy density does not vanish at some points in space. The distribution of energy density is the topological defect which, depending on dimensionality, can manifest itself as a domain wall ($n = 0$), string ($n = 1$) or monopole ($n = 2$) or texture ($n = 3$).

The above argument establishes that topologically non-trivial boundary conditions imply non-vanishing energy in the field. However, it does not establish that a static solution exists with those boundary conditions. These must be found on a case-by-case basis. Indeed there are examples of topologically non-trivial boundary conditions where no static solution exists.[3] Also distinct elements of $\pi_n(M)$ ($n \geq 1$) need not lead to distinct field solutions. Only those solutions that correspond to elements of $\pi_n(M)$ that cannot be continuously deformed into each other, *if the maps are released from the base point*, are distinct. The italicized remark is in recognition of the fact that there can be two maps that are mathematically distinct (i.e. cannot be deformed into each other) only because they are fixed at the base point. However, the analog of a "common base point" in field theory would be to restrict attention to field configurations for which the fields attain a certain fixed value at some point on S_∞^n. Such a restriction is generally unphysical and hence, we are interested in

---

[3] For example, in three dimensions, the boundary conditions corresponding to a charge two 't Hooft-Polyakov magnetic monopole [124, 79] do not lead to any solution (for all but one value of model parameters). This is because any field configuration with those boundary conditions breaks up into two magnetic monopoles, each of unit charge, that repel each other and are never static.

maps that cannot be deformed into each other even if we release the restriction that all maps have a common base point (for a more detailed discussion, see [171]).

In the case when the vacuum manifold has disconnected components, $\pi_0(G/H)$ is non-trivial since there are points (zero-dimensional spheres) that lie in different components that cannot be continuously deformed into one another. Therefore kinks occur whenever $\pi_0(G/H)$ is non-trivial. In the $\lambda\Phi^4$ model, $G = Z_2$, $H = 1$ and $\pi_0(G/H) = Z_2$. In the sine-Gordon model $G = Z$, $H = 1$ and $\pi_0(G/H) = Z$. If $\pi_0 = Z_N$, we name the resulting kinks "Z_N kinks." In these simple examples, π_0 forms a group because G is Abelian and so G/H itself is a group. An example in which π_0 is not a group can be constructed by choosing $G = S_3$ (S_n is the permutation group of n elements) broken down to $H = S_2$.

The kinks in a model with disconnected elements in M can now be classified. Every element of $\pi_0(M)$ corresponds to a mapping from a point at spatial infinity to M and hence specifies a domain at infinity. Kinks occur if the domains at $\pm\infty$ are distinct. Therefore pairs of elements of $\pi_0(M)$ classify domain walls.

1.11 Bogomolnyi method revisited

The Bogomolnyi method can be extended to include a large class of systems. Let us start with the general energy functional for a matrix-valued complex scalar field Φ

$$E = \int dx \left[\mathrm{Tr}|\partial_t\Phi|^2 + \mathrm{Tr}|\partial_x\Phi|^2 + V(\Phi, \Phi^*) \right]$$
$$= \int dx \left[\partial_t\Phi^*_{ab}\partial_t\Phi_{ba} + \partial_x\Phi^*_{ab}\partial_x\Phi_{ba} + V(\Phi, \Phi^*) \right] \quad (1.64)$$

where a sum over matrix components labeled by a, b is understood. As in Section 1.5, we would like to write the energy density in "whole square" form

$$E = \int dx \left[\mathrm{Tr}\{|\partial_t\Phi|^2 + |\partial_x\Phi \mp U(\Phi)|^2 \pm (\partial_x\Phi^\dagger U) \pm (U^\dagger\partial_x\Phi)\} \right] \quad (1.65)$$

where we are restricting ourselves to static solutions and U is some matrix-valued function of Φ such that

$$\mathrm{Tr}(U^\dagger U) = V(\Phi, \Phi^*) \quad (1.66)$$

The energy is minimized if

$$\partial_t\Phi = 0 \quad (1.67)$$

and

$$\mathrm{Tr}|\partial_x\Phi \mp U(\Phi, \Phi^*)|^2 = 0 \quad (1.68)$$

which in turn gives

$$\partial_x \Phi \mp U(\Phi, \Phi^*) = 0 \tag{1.69}$$

The energy of the kink is

$$E = \pm \int_{-\infty}^{+\infty} dx \, \mathrm{Tr}(\partial_x \Phi^\dagger U + U^\dagger \partial_x \Phi) \tag{1.70}$$

There is a further special case – the "supersymmetric" case – in which the energy integral can be performed explicitly. This is if U is a total derivative

$$U^* = \frac{\partial W}{\partial \Phi} \tag{1.71}$$

where $W(\Phi, \Phi^*)$ is the "superpotential," assumed to be real. Then

$$\begin{aligned}
E &= \pm \int_{-\infty}^{+\infty} dx \, \mathrm{Tr} \left(\partial_x \Phi^\dagger \frac{\partial W}{\partial \Phi^*} + \partial_x \Phi^T \frac{\partial W}{\partial \Phi} \right) \\
&= \pm \int_{-\infty}^{+\infty} dx \, \partial_x W \\
&= \pm [W(\Phi(+\infty)) - W(\Phi(-\infty))]
\end{aligned} \tag{1.72}$$

Therefore we see that the Bogomolnyi method allows for first-order equations of motion provided that V can be written as $\mathrm{Tr}(U^\dagger U)$. The method also provides an explicit expression for the kink energy if V is given in terms of a superpotential W as

$$V(\Phi) = \mathrm{Tr}(U^\dagger U) = \mathrm{Tr} \left| \frac{dW}{d\Phi} \right|^2 \tag{1.73}$$

1.12 On more techniques

The kink solutions we have been discussing fall under the more general category of "solitary waves," often discussed under the soliton heading. Strictly speaking, for a solution to classify as a "soliton," it also has to satisfy certain conditions on its scattering with other solitons. The subject is incredibly rich, and has led to the development of very sophisticated mathematical techniques such as Backlund transformations, inverse scattering methods, Lax heirarchy, etc. In addition, solitons have found tremendous importance in physical applications, especially non-linear optics and communication. Readers interested in the mathematics and physics of solitons might wish to consult [1, 48, 56].

Strict solitons are usually discussed in one spatial dimension and have limited application in the context of particle physics. Nonetheless, there are equally sophisticated techniques to study solitary wave solutions in higher dimensions. In particular, the ADHM construction [12] is used to find instanton solutions in four spatial dimensions and the Nahm equations lead to magnetic monopole solutions in three dimensions [114].

The soliton analyses mentioned above consider equations with complicated non-linear terms and higher derivatives. In the context of particle physics, such terms and derivatives are rarely encountered. However, one complication that arises is due to larger (non-Abelian) symmetry groups. In the next chapter we will take the analysis of this section to such particle-physics motivated models. There we will find a spectrum of kink solutions with unusual interactions. As we proceed to further chapters, we will learn that the physics of such non-Abelian kinks can be quite different from that of the simple kinks discussed in this chapter.

1.13 Open questions

1. Discuss the conditions needed for a breather solution to exist. If an exact breather does not exist, can there be an approximate breather (see Section 3.1)? What is the approximation?

2

Kinks in more complicated models

The Z_2 and sine-Gordon kinks discussed in the last chapter are not representative of kinks in models where non-Abelian symmetries are present. Kinks in such models have more degrees of freedom and this introduces degeneracies when imposing boundary conditions, leading to many kink solutions with different internal structures (but the same topology). Indeed, kink-like solutions may exist even when the topological charge is zero. The interactions of kinks in these more complicated models, their formation and evolution, plus their interactions with other particles are very distinct from the kinks of the last chapter.

We choose to focus on kinks in a model that is an example relevant to particle physics and cosmology. The model is the first of many Grand Unified Theories of particle physics that have been proposed [63]. The idea behind grand unification is that Nature really has only one gauge-coupling constant at high energies, and that the disparate values of the strong, weak, and electromagnetic coupling constants observed today are due to symmetry breaking and the renormalization-group running of coupling constants down to low energies. Since there is only one gauge-coupling constant in these models, there is a simple grand unified symmetry group G that is valid at high energies, for example, at the high temperatures present in the very early universe. At lower energies, G is spontaneously broken in stages, eventually leaving only the presently known quantum chromo dynamics (QCD) and electromagnetic symmetries $SU(3)_c \times U(1)_{em}$ of particle physics, with its two different coupling constants. It can be shown [63] that the minimal possibility for G is $SU(5)$. However, since Grand Unified Theories predict proton decay, experimental observation of the longevity of the proton ($\sim 5 \times 10^{33}$ years) leads to constraints on grand unified models. The (non-supersymmetric) $SU(5)$ Grand Unified Theory is ruled out by the current lower limits on the proton's lifetime. Therefore particle-physics model builders consider yet larger groups G, or with an extended scalar field sector, or supersymmetric extensions of $SU(5)$, and other models based on larger groups. Even if the symmetry group is larger than $SU(5)$, it often happens that after a series of symmetry breaking, the residual symmetry is $SU(5)$, which then proceeds to break to the current symmetry group. Hence the study of $SU(5)$ symmetry breaking is extremely relevant to particle physics, even if it is not the ultimate grand unified symmetry group.

In this chapter we shall study kinks in a model with $SU(5) \times Z_2$ symmetry though almost all the discussion can be generalized to an $SU(N) \times Z_2$ model for odd values of N [163, 120]. The extra Z_2 symmetry is explained in the next section. Since we only desire to study kinks in a particle-physics motivated model, it would seem simpler to choose a model based on the smaller $SU(3)$ group. However, it can be shown that there is no way to construct a model with just $SU(3)$ symmetry and with the simplest choice of field content, which is one adjoint field. Instead, the model must have the larger $O(8)$ symmetry. Other fields need to be included so as to reduce the $O(8)$ to $SU(3)$, but that introduces additional parameters which make the $SU(3)$ model more messy than the $SU(5)$ model.

Dealing with continuous groups such as $SU(5)$ requires certain background material. The fundamental representation of $SU(N)$ generators is described in Appendix B. A summary of some aspects of the $SU(5)$ model of grand unification is given in Section 5.5.

2.1 $SU(5)$ model

The $SU(5)$ model can be written as[1]

$$L = \text{Tr}(D_\mu \Phi)^2 - \frac{1}{2}\text{Tr}(X_{\mu\nu} X^{\mu\nu}) - V(\Phi) \tag{2.1}$$

where, in terms of components, Φ is a scalar field (also called a Higgs field) transforming in the adjoint representation of $SU(5)$, that is, $\Phi \rightarrow \Phi' = g\Phi g^\dagger$ for $g \in SU(5)$. The gauge field strengths are $X_{\mu\nu} = X_{\mu\nu}^a T^a$ and the $SU(5)$ generators T^a are normalized such that $\text{Tr}(T^a T^b) = \delta^{ab}/2$. The definition of the covariant derivative is

$$D_\mu = \partial_\mu - \text{ie}X_\mu \tag{2.2}$$

and its action on the adjoint scalar is given by

$$D_\mu \Phi = \partial_\mu \Phi - \text{ie}[X_\mu, \Phi] \tag{2.3}$$

The gauge field strength is given in terms of the covariant derivative via

$$-\text{ie}X_{\mu\nu} = [D_\mu, D_\nu] \tag{2.4}$$

---

[1] We are using the Einstein summation convention in which repeated group and space-time indices are summed over. So, explicitly, $\Phi = \sum_{a=1}^{24} \Phi^a T^a$. See Appendix B for more details on the $SU(5)$ generators T^a.

and the potential is the most general quartic in Φ

$$V(\Phi) = -m^2 \text{Tr}(\Phi^2) + h[\text{Tr}(\Phi^2)]^2 + \lambda \text{Tr}(\Phi^4) + \gamma \text{Tr}(\Phi^3) - V_0 \qquad (2.5)$$

where V_0 is a constant that is chosen so as to set the minimum value of the potential to zero.

The model in Eq. (2.1) does not have any topological kinks because there are no broken discrete symmetries. In particular, the Z_2 symmetry under $\Phi \to -\Phi$ is absent owing to the cubic term in Eq. (2.5). Note that $\Phi \to -\Phi$ is not achievable by an $SU(5)$ transformation. To show this, consider $\text{Tr}(\Phi^3)$. This is invariant under any $SU(5)$ transformation, but not under $\Phi \to -\Phi$. However, if $\gamma = 0$, there are topological kinks connecting the two vacua related by $\Phi \to -\Phi$. For non-zero but small γ, these kinks are almost topological. In our analysis in this chapter we set $\gamma = 0$, in which case the symmetry of the model is $SU(5) \times Z_2$. The philosophy underlying grand unification does not forbid discrete symmetry factors since such factors do not entail additional gauge-coupling constants. Indeed, model builders often set $\gamma = 0$ for simplicity. Now a non-zero vacuum expectation value of Φ breaks the discrete Z_2 factor leading to topological kinks.

2.2 $SU(5) \times Z_2$ symmetry breaking and topological kinks

The potential in Eq. (2.5) has a (degenerate) global minimum at

$$\Phi_0 = \frac{\eta}{2\sqrt{15}} \text{diag}(2, 2, 2, -3, -3) \qquad (2.6)$$

where $\eta = m/\sqrt{\lambda'}$ provided

$$\lambda \geq 0, \qquad \lambda' \equiv h + \frac{7}{30}\lambda \geq 0 \qquad (2.7)$$

For the global minimum to have $V(\Phi_0) = 0$, in Eq. (2.5) we set

$$V_0 = -\frac{\lambda'}{4}\eta^4 \qquad (2.8)$$

As discussed in Section 1.10, if we transform Φ_0 by any element of $SU(5) \times Z_2$, the transformed Φ_0 is still at a minimum of the potential. However, Φ_0 is left unmoved by transformations belonging to

$$G_{321} \equiv \frac{[SU(3) \times SU(2) \times U(1)]}{Z_3 \times Z_2} \qquad (2.9)$$

where $SU(3)$ acts on the upper-left 3×3 block of Φ_0, $SU(2)$ on the lower-right 2×2 block, and $U(1)$ is generated by Φ_0 itself. Hence, G_{321} is the unbroken symmetry group.

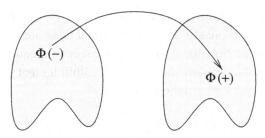

Figure 2.1 The vacuum manifold of the $SU(5) \times Z_2$ model consists of two disconnected 12-dimensional copies. Kink solutions correspond to paths that originate in one piece at $x = -\infty$, denoted by $\Phi(-)$, leave the vacuum manifold, and end in the other disconnected piece at $x = +\infty$. Topological considerations specify that $\Phi(+)$ has to lie in the disconnected piece on the right, but not where it should be located within this piece.

$SU(5)$ has 24 generators while the unbroken group, G_{321}, has a total of 12 generators, namely, 8 of $SU(3)$, 3 of $SU(2)$, and 1 of $U(1)$. Therefore the vacuum manifold is $24 - 12 = 12$ dimensional but in two disconnected pieces as depicted in Fig. 2.1 because of the Z_2 factor. Kink solutions occur if the boundary conditions lie in different disconnected pieces. However, if we start at some point on the vacuum manifold at $x = -\infty$, say $\Phi(-\infty) = \Phi_-$, we have a choice of boundary conditions for Φ_+, the vacuum expectation value of Φ at $x = +\infty$ (compare with the Z_2 case where the path had to go from definite initial to definite final values of Φ).

We will narrow down the possible choices for Φ_+ very shortly. First we point out that the gauge fields can be set to zero in finding kink solutions [163]. To see this explicitly, the only linear term in the gauge field is ieTr($X^i[\Phi, \partial_i\Phi]$). However, our solution for Φ satisfies $[\Phi, \partial_i\Phi] = 0$ [120] and so the variation vanishes to linear order in gauge field fluctuations. A closer look also reveals that the quadratic terms of perturbations in the gauge fields contribute positively to the energy of the kink solutions and so the gauge fields do not cause an instability of the solutions [163]. Hence we set

$$X_\mu = 0 \tag{2.10}$$

As we now show, the boundary conditions that lead to static solutions of the equations of motion are rather special [120].

Theorem: A static solution can exist only if $[\Phi_+, \Phi_-] = 0$.

We only give a sketch of the proof here since it is of a technical nature. The essential idea is that if $\Phi_k(x)$ is a static solution, then the energy should be extremized by it. By considering perturbations of the kind $U(x)\Phi_k U^\dagger(x)$ where $U(x)$ is an infinitesimal rotation of $SU(5)$, one finds that the energy can be extremized only if

$[\Phi_k, \partial_x \Phi_k] = 0$ for all x. Now at large x, we have $\Phi_k \rightarrow \Phi_+$. In this region $\partial_x \Phi_k$ has terms that are proportional to Φ_- as well, even if these are exponentially small, since $\Phi(x)$ is an analytic function. Hence, a static solution requires $[\Phi_+, \Phi_-] = 0$.

The theorem immediately narrows down the possibilities that we need to consider when trying to construct kink solutions. If we fix

$$\Phi_- = \Phi_0 = \frac{\eta}{2\sqrt{15}} \text{diag}(2, 2, 2, -3, -3) \tag{2.11}$$

Φ_+ can take on the following three values

$$\Phi_+^{(0)} = -\frac{\eta}{2\sqrt{15}} \text{diag}(2, 2, 2, -3, -3)$$

$$\Phi_+^{(1)} = -\frac{\eta}{2\sqrt{15}} \text{diag}(2, 2, -3, 2, -3)$$

$$\Phi_+^{(2)} = -\frac{\eta}{2\sqrt{15}} \text{diag}(2, -3, -3, 2, 2) \tag{2.12}$$

One can also rotate these three choices by elements of the unbroken group G_{321-} that leaves Φ_- invariant and obtain three disjoint classes of possible values of Φ_+. The three choices given above are representatives of their classes.

The kink solution for any of the three boundary conditions is of the form

$$\phi_k^{(q)} = F_+^{(q)}(x)M_+^{(q)} + F_-^{(q)}(x)M_-^{(q)} + g^{(q)}(x)M^{(q)} \tag{2.13}$$

where $q = 0, 1, 2$ labels the solution class,

$$M_+^{(q)} = \frac{\Phi_+^{(q)} + \Phi_-^{(q)}}{2}, \qquad M_-^{(q)} = \frac{\Phi_+^{(q)} - \Phi_-^{(q)}}{2} \tag{2.14}$$

and $M^{(q)}$ will be specified below.

The boundary conditions for $F_\pm^{(q)}$ are

$$F_-^{(q)}(\mp\infty) = \mp 1, \qquad F_+^{(q)}(\mp\infty) = +1, \qquad g^{(q)}(\mp\infty) = 0 \tag{2.15}$$

The formulae for $M_\pm^{(q)}$ and $M^{(q)}$ can now be explicitly written using Eq. (2.12) in (2.14)

$$M_+^{(q)} = \eta \frac{5}{4\sqrt{15}} \text{diag}(0_{3-q}, 1_q, -1_q, 0_{2-q}) \tag{2.16}$$

$$M_-^{(q)} = \eta \frac{1}{4\sqrt{15}} \text{diag}(-4 1_{3-q}, 1_q, 1_q, 6 1_{2-q}) \tag{2.17}$$

$$M^{(q)} = \mu \, \text{diag}(q(2-q)1_{3-q}, -(2-q)(3-q)1_{2q}, q(3-q)1_{2-q}) \tag{2.18}$$

with the normalization μ given by

$$\mu = \eta[2q(2-q)(3-q)(12-5q)]^{-1/2} \tag{2.19}$$

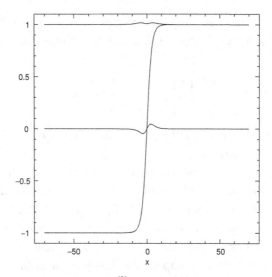

Figure 2.2   The profile functions $F_+^{(1)}(x)$ (nearly 1 throughout), $F_-^{(1)}(x)$ (shaped like a tanh function), and $g^{(1)}(x)$ (nearly zero) for the $q = 1$ topological kink with parameters $h = -3/70$, $\lambda = 1$, and $\eta = 1$.

If $q = 0$ or $q = 2$ we set $\mu = 0$. We have used $\mathbf{0}_k$ and $\mathbf{1}_k$ to denote the $k \times k$ zero and unit matrices respectively. Note that the matrices $\mathbf{M}_\pm^{(q)}$ are relatively orthogonal

$$\text{Tr}(\mathbf{M}_+^{(q)}\mathbf{M}_-^{(q)}) = 0 \qquad (2.20)$$

but are not normalized to $\eta^2/2$.

Now we discuss the three kink solutions in the $SU(5) \times Z_2$ model. For $q = 0$, the solution is that of a Z_2 kink that has been embedded in the $SU(5) \times Z_2$ model. The explicit solution is

$$F_+^{(0)}(x) = 0, \qquad F_-^{(0)}(x) = -\tanh\left(\frac{x}{w}\right), \qquad g^{(0)}(x) = 0 \qquad (2.21)$$

where $w = \sqrt{2}/m$. For $q = 1$, the profile functions have been evaluated numerically and are shown in Fig. 2.2. Approximate analytic solutions can also be found in [120]. For $q = 2$ the solution has also been found numerically. Here we describe an approximate solution which is exact if

$$\frac{h}{\lambda} = -\frac{3}{20} \qquad (2.22)$$

i.e. $\lambda' = \lambda/12$. With this particular choice

$$F_+^{(2)}(x) = 1, \qquad F_-^{(2)}(x) = \tanh\left(\frac{x}{w}\right), \qquad g^{(2)}(x) = 0 \qquad (2.23)$$

where $w = \sqrt{2}/m$. This is also an approximate solution for $h/\lambda \approx -3/20$. The energy of the approximate solution can be used to estimate the mass of the $q = 2$ kink

$$M^{(2)} \approx \frac{M^{(0)}}{6} \left\{ \frac{1}{6} \left[1 + \frac{5\lambda}{12\lambda'} \right] \right\}^{1/2} \equiv M^{(0)} \frac{\sqrt{p}}{6} \qquad (2.24)$$

where $M^{(2)}$ denotes the mass of the $q = 2$ kink, and $M^{(0)} = 2\sqrt{2}m^3/3\lambda'$. The expression for the energy is exact for $h/\lambda = -3/20$.

It can be shown for a range of parameters that the $q = 2$ kink solution is perturbatively stable. Numerical evaluations of the energy find that the $q = 2$ kink is lighter than the $q = 0, 1$ kinks for all values of p. Equation (2.24) shows the $q = 2$ kink is lighter than the $q = 0$ kink for a large range of parameters. This can be understood qualitatively by noting that only one component of Φ changes sign in the $q = 2$ kink, while 3 and 5 components change sign in the $q = 1$ and $q = 0$ kinks respectively.

2.3 Non-topological $SU(5) \times Z_2$ kinks

An interesting point to note is that the ansatz in Eq. (2.13) is valid even if $\Phi_{\pm}^{(q)}$ are not in distinct topological sectors. These imply the existence of non-topological kink solutions in the model [120]. If we include a subscript NT to denote "non-topological" and T to denote "topological," we have

$$\Phi_{\text{NT}k}^{(q)} = F_+^{(q)}(x)\mathbf{M}_{\text{NT}+}^{(q)} + F_-^{(q)}(x)\mathbf{M}_{\text{NT}-}^{(q)} + g^{(q)}(x)\mathbf{M}_{\text{NT}}^{(q)} \qquad (2.25)$$

where the $\mathbf{M}_{\text{NT}\pm}$ matrices are still defined by Eq. (2.14) with the non-topological values of $\Phi_{\pm}$. $\mathbf{M}_{\text{NT}}$ is still given by Eq. (2.18). To consider a non-topological domain wall, we simply want to consider Φ_+ to be in the same discrete sector as Φ_-. If $\Phi_{\text{T}+}$ denotes a boundary condition for a topological kink, a possible boundary condition for a non-topological kink is: $\Phi_{\text{NT}+} = -\Phi_{\text{T}+}$. Then we find

$$\mathbf{M}_{\text{NT}+}^{(q)} = \mathbf{M}_{\text{T}-}^{(q)}, \qquad \mathbf{M}_{\text{NT}-}^{(q)} = \mathbf{M}_{\text{T}+}^{(q)}, \qquad \mathbf{M}_{\text{NT}}^{(q)} = \mathbf{M}_{\text{T}}^{(q)} \qquad (2.26)$$

Hence

$$\Phi_{\text{NT}k}^{(q)} = F_-^{(q)}(x)\mathbf{M}_{\text{T}+}^{(q)} + F_+^{(q)}(x)\mathbf{M}_{\text{T}-}^{(q)} + g^{(q)}(x)\mathbf{M}_{\text{T}}^{(q)} \qquad (2.27)$$

To get $F_{\mp}^{(q)}$ for the non-topological kink we have to solve the topological $F_{\pm}^{(q)}$ equation of motion but with the boundary conditions for $F_{\mp}^{(q)}$ (see Eq. (2.15)). To obtain $g^{(q)}$ for the non-topological kink, we need to interchange $F_+^{(q)}$ and $F_-^{(q)}$ in the topological equation of motion. The boundary conditions for $g^{(q)}$ are unchanged. Generally the non-topological solutions, when they exist, are unstable. However,

Table 2.1 *The space of three topological kinks in the SU(5) model.*
G_{321} is the group $SU(3) \times SU(2) \times U(1)$. The dimensionality of the space
of each type of kink is also given.

Kink	Space	Dimensionality
$q = 0$	G_{321}/G_{321}	0
$q = 1$	$G_{321}/[SU(2) \times U(1)^3]$	6
$q = 2$	$G_{321}/[SU(2)^2 \times U(1)^2]$	4

the possibility that some of them may be locally stable for certain potentials cannot
be excluded.

2.4 Space of $SU(5) \times Z_2$ kinks

The kink solutions discussed in Section 2.1 can be transformed into other degenerate
solutions using the $SU(5)$ transformations. Hence, each solution is representative
of a space of solutions. We now discuss the space associated with each of these
solutions.

If we denote a kink solution in the $SU(5) \times Z_2$ model by $\Phi_k^{(q)}$, another solution is

$$\phi_k^{(q)h} = h\phi_k^{(q)}h^\dagger, \quad h \in G_{321-} \tag{2.28}$$

where G_{321-} is the unbroken group whose elements leave Φ_- unchanged.[2] The
reason $\Phi_k^{(q)h}$ also describes a solution is that the rotation h does not change the
energy of the field configuration, $\Phi_k^{(q)}$. Therefore $\Phi_k^{(q)h}$ has the same energy and
the same topology as $\Phi_k^{(q)}$, and hence it describes another kink solution.

Of the elements of G_{321-}, there are some that act trivially on $\Phi_k^{(q)}$ and for these
h, $\Phi_k^{(q)h}$ is not distinct from $\Phi_k^{(q)}$. These elements form a subgroup of G_{321-} that we
call K_q. Therefore the space of kinks can be labeled by elements of the coset space
G_{321-}/K_q. Since we are given the forms of the kink solutions in Eq. (2.13), it is not
hard to work out K_q. For example, for the $q = 2$ kink, K_q is given by the $SU(5)$
elements that commute with both G_{321-} and G_{321+} and so $K_q = SU(2)^2 \times U(1)^2$.
Once we have determined K_q the dimensionality of the coset space G_{321-}/K_q is
determined as the dimensionality of G_{321-}, which is 12, minus the dimensionality
of K_q, which is 12, 6, and 8 for $q = 0$, 1, and 2 respectively.

The three classes of kink solutions labeled by the index q in the $SU(5) \times Z_2$
model have different spaces as shown in Table 2.1.

---

[2] We could also have included elements that change $\Phi_+^{(q)}$ as well as Φ_-. These would simply be global rotations
of the entire solution and would be the same for every type of defect.

The dimensionality of the space of a given type of kink solution also corresponds to the dimensionality of the space of boundary conditions Φ_+ for which that type of kink solution is obtained. As an example, there is only one value of Φ_+, namely $\Phi_+ = -\Phi_-$, that gives rise to the $q = 0$ kink. While for the $q = 1$ kink, one can choose Φ_+ to be any value from a 6-dimensional space. This means that, in any process where boundary conditions are chosen at random, the probabilities of getting the correct boundary conditions for a $q = 0$ or a $q = 2$ kink are of measure zero, since the space of boundary conditions for the $q = 1$ kink is two dimensions greater than that for the $q = 2$ kink. In any random process, the $q = 1$ kink is always obtained. Since this kink is unstable, it then decays into the $q = 2$ kink. Therefore the production of $q = 2$ kinks is a two-step process in this system. We will see further evidence of this two-step process in Chapter 6.

2.5 S_n kinks

The $SU(5) \times Z_2$ model discussed above shows novel features because of the large non-Abelian symmetry. It is possible to see some of the richness of the model by going to a simpler model where the continuous non-Abelian symmetries are replaced by discrete non-Abelian symmetries (also see [92] for a similar model). If we truncate the $SU(5) \times Z_2$ model to just the diagonal degrees of freedom of Φ, we get a model that is symmetric only under permutations of the diagonal entries and the overall Z_2. Hence the symmetry group is $S_5 \times Z_2$, where S_5 is the permutation group of five objects. The model now has four real scalar fields, one for each diagonal generator of $SU(5)$. With this truncation we can write

$$\Phi \rightarrow f_1 \lambda_3 + f_2 \lambda_8 + f_3 \tau_3 + f_4 Y \tag{2.29}$$

where the f_i are functions of space and time, and the generators $\lambda_3, \lambda_8, \tau_3$, and Y are defined in Appendix B. Inserting this form of Φ into the $SU(5) \times Z_2$ Lagrangian in Eq. (2.1) we get

$$L = \frac{1}{2} \sum_{i=1}^{4} (\partial_\mu f_i)^2 + V(f_1, f_2, f_3, f_4) \tag{2.30}$$

and

$$V = -\frac{m^2}{2} \sum_{i=1}^{4} f_i^2 + \frac{h}{4} \left(\sum_{i=1}^{4} f_i^2 \right)^2 + \frac{\lambda}{8} \sum_{a=1}^{3} f_a^4 + \frac{\lambda}{4} \left[\frac{7}{30} f_4^4 + f_1^2 f_2^2 \right]$$

$$+ \frac{\lambda}{20} [4(f_1^2 + f_2^2) + 9f_3^2] f_4^2 + \frac{\lambda}{\sqrt{5}} f_2 f_4 \left(f_1^2 - \frac{f_2^2}{3} \right) + \frac{m^2}{4} \eta^2 \tag{2.31}$$

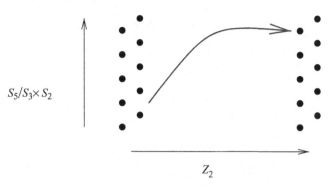

$S_5/S_3 \times S_2$

Z_2

Figure 2.3 The vacuum manifold for the $S_5 \times Z_2$ model contains two sets of ten points related by the Z_2 symmetry. Kink solutions exist that interpolate between vacua related by Z_2 transformations and also between vacua within one set of ten points. The former correspond to the topological kinks in $SU(5) \times Z_2$ and the latter to the non-topological kinks in that model.

This model has the desired $S_5 \times Z_2$ symmetry because it is invariant under permutations of the diagonal elements of Φ, that is, under permutations of various linear combinations of f_i. The Z_2 symmetry is under $f_i \to -f_i$ for every i.

Symmetry breaking proceeds as in the $SU(5) \times Z_2$ case. The $S_5 \times Z_2$ symmetry is broken by a vacuum expectation value along the Y direction i.e. $f_4 \neq 0$. The residual symmetry group consists of permutations in the $SU(3)$ and $SU(2)$ blocks. Therefore the unbroken symmetry group is $H = S_3 \times S_2$. There are $5! \times 2 = 240$ elements of $S_5 \times Z_2$ and $3! \times 2! = 12$ elements of H. Therefore the vacuum manifold consists of $240/12 = 20$ distinct points. Ten of these points are related to the other ten by the non-trivial element of Z_2 as shown in Fig. 2.3. If we fix the boundary condition at $x = -\infty$, then a Z_2 kink can be obtained with ten different boundary conditions at $x = +\infty$. These ten solutions must somehow correspond to the kink solutions that we have already found in the $SU(5) \times Z_2$ case. Counting all the possible different diagonal possibilities for Φ_+ in the $SU(5) \times Z_2$ model we see that there are three $q = 2$ kinks, six $q = 1$ kinks, and one $q = 0$ kink, making a total of ten kinks. In the $S_5 \times Z_2$ model there are ten more (one of these is the trivial solution) kinks that do not involve the Z_2 transformation (change of sign) in going from Φ_- to Φ_+. These are the ten remnants of the non-topological kinks described in Section 2.3.

2.6 Symmetries within kinks

The symmetry groups outside the kink, $G_{321\pm}$, are isomorphic (see Fig. 2.4). However, the fields transform differently under the elements of these groups. As a result, there is a "clash of symmetries" [43] inside the kink, and the unbroken symmetry

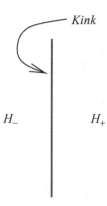

Figure 2.4   A kink and the symmetries outside denoted by $H_\pm$. The groups H_+ and H_- are isomorphic but their action on fields may not necessarily be identical.

group within the kink is generally *smaller* than that outside. This does not happen in the case of the Z_2 kink in which the symmetry outside is trivial while inside it is Z_2 (since the field vanishes). We now examine the clash of symmetries in the case of the $SU(5) \times Z_2\, q = 2$ kink.

The general form of $\Phi_k^{(2)}$ is given in Eq. (2.13) with the profile functions in Eq. (2.23). Then

$$\Phi_k^{(2)}(x = 0) = M_+^{(2)} \propto \text{diag}(0, 1, 1, -1, -1) \tag{2.32}$$

The symmetries within the kink are given by the elements of $SU(5) \times Z_2$ that leave $M_+^{(2)}$ invariant. Hence the internal symmetry group consists of two $SU(2)$ factors, one for each block proportional to the 2×2 identity, and two $U(1)$ factors since all diagonal elements of $SU(5)$ commute with $M_+^{(2)}$. Therefore the symmetry group inside the $SU(5) \times Z_2$ kink is $[SU(2)]^2 \times [U(1)]^2$. This is *smaller* than the $SU(3) \times SU(2) \times U(1)$ symmetry group outside the kink.[3]

The conclusion that the symmetry inside a kink is smaller than that outside holds quite generally [164]. Classically this would imply that there are more massless particles outside the kink than inside it. However, when quantum effects are taken into account this classical picture can change because the fundamental states in the outside region may consist of confined groups of particles ("mesons" and "hadrons") that are very massive [51]. If a particle carries non-Abelian charge of a symmetry that is unbroken outside the wall but broken inside to an Abelian subgroup, it may cost less energy for the particle to live on the wall. This is because it may be

---

[3] As in Section 2.4 we could have found the symmetry group inside the kink by finding those transformations in G_{321-} that are also contained in G_{321+}.

unconfined inside the wall where it only carries Abelian charge, while it can only exist as a heavy meson or a hadron outside the wall.[4]

2.7 Interactions of static kinks in non-Abelian models

The interaction potential between kinks found in Section 1.8 is easily generalized to kinks in non-Abelian field theories. Following the procedure discussed in that section, the force in the $SU(5) \times Z_2$ case is

$$F = \frac{dP}{dt} = \left[- \text{Tr}(\dot{\Phi}^2) - \text{Tr}(\Phi'^2) + V(\Phi) \right]_{x_1}^{x_2} \qquad (2.33)$$

where $-a - R$ and $-a + R$ are defined in Fig. 1.3. Evaluation of F yields an exponentially small interaction force whose sign depends on $\text{Tr}(Q_1 Q_2)$ [121] where Q_1 and Q_2 are the topological charges of the kinks. If the Higgs field at $x = -\infty$ is Φ_-, between the two kinks is Φ_0, and is Φ_+ at $x = +\infty$, then $Q_1 \propto \Phi_0 - \Phi_-$ and $Q_2 \propto \Phi_+ - \Phi_0$ (see Eq. (1.8)).

What is most interesting about the interaction is that a kink and an antikink can repel. Here one needs to be careful about the meaning of an "antikink." An antikink should have a topological charge that is opposite to that of a kink. That is, a kink and its antikink together should be in the trivial topological sector. But this condition still leaves open several different kinds of antikinks for a given kink. To be specific consider a kink-antikink pair, where the Higgs field across the kink changes from $\Phi(-\infty) \propto +(2, 2, 2, -3, -3)$ to $\Phi(0) \propto -(2, -3, -3, 2, 2)$. (Here we suppress the normalization factor and the "diag" for convenience of writing.) There can be two types of antikinks to the right of this kink. In the first type (called Type I) the Higgs field can go from $\Phi(0) \propto -(2, -3, -3, 2, 2)$ to $\Phi(+\infty) \propto +(2, 2, 2, -3, -3)$, which is the same as the value of the Higgs field at $x = -\infty$ and thus reverts the change in the Higgs across the kink. In the second type (Type II), the Higgs field can go from $\Phi(0) \propto -(2, -3, -3, 2, 2)$ to $\Phi(+\infty) \propto +(-3, 2, 2, -3, 2)$. Now the Higgs at $x = +\infty$ is not the same as the Higgs at $x = -\infty$, but the two asymptotic field values are in the same topological sector.

By evaluating $\text{Tr}(Q_1 Q_2)$, where Q_1 and Q_2 are the charge matrices of the two kinks, it is easy to check that the force between a kink and its Type I antikink is attractive, but the force between a kink and its Type II antikink is repulsive. The $q = 2$ kinks can have charge matrices $Q^{(i)}$ that we list up to a proportionality factor

$$Q^{(1)} = (-4, 1, 1, 1, 1), \qquad Q^{(2)} = (1, -4, 1, 1, 1), \qquad Q^{(3)} = (1, 1, -4, 1, 1),$$
$$Q^{(4)} = (1, 1, 1, -4, 1), \qquad Q^{(5)} = (1, 1, 1, 1, -4) \qquad (2.34)$$

---

[4] Localization of particles to the interior of defects has led to the construction of cosmological scenarios where our observed universe is a three-dimensional defect or "brane" embedded in a higher dimensional space-time.

Stable antikinks have the same charges but with a minus sign. Then, one can take a kink with one of the five charges listed above and it repels an antikink that has the -4 occurring in a different entry because $\text{Tr}(Q_1 Q_2) > 0$. Hence, there are combinations of kinks and antikinks for which the interaction is repulsive. Further, in a statistical system a kink is most likely to have a Type II antikink as a neighbor and such a kink-antikink pair cannot annihilate since the force is repulsive.

The result that the force between two kinks is proportional to the trace of the product of the charges extends to other solitons (e.g. magnetic monopoles) as well. In this way, the forces between certain monopoles with equivalent magnetic charge can be attractive whereas normally we would think that like magnetic charges repel, and between certain monopoles and antimonopoles can be repulsive.

2.8 Kink lattices

In this section we describe the possibility of forming stable lattices of domain walls in one spatial dimension and the consequences in higher dimensions. Our discussion is in the context of the $S_5 \times Z_2$ model though similar structures have been seen in other field theory models as well [92, 43].

We know that Z_2 topology forces a kink to be followed by an antikink. Then we can set up a sequence of kinks and antikinks whose charges are arranged in the following way

$$\ldots Q^{(1)} \bar{Q}^{(5)} Q^{(3)} \bar{Q}^{(1)} Q^{(5)} \bar{Q}^{(3)} \ldots \tag{2.35}$$

where $Q^{(i)}$ and $\bar{Q}^{(i)}$ refer to a kink and an antikink of type i respectively (see Eq. (2.34)). Alternately, this sequence of kinks would be achieved with the following sequence of Higgs field vacuum expectation values (illustrated in Fig. 2.5)

$$
\begin{aligned}
\ldots &\rightarrow -(2, 2, 2, -3, -3) \rightarrow +(2, -3, -3, 2, 2) \\
&\rightarrow -(-3, 2, 2, -3, 2) \\
&\rightarrow +(2, -3, 2, 2, -3) \\
&\rightarrow -(2, 2, -3, -3, 2) \\
&\rightarrow +(-3, -3, 2, 2, 2) \\
&\rightarrow -(2, 2, 2, -3, -3) \rightarrow \ldots
\end{aligned}
\tag{2.36}
$$

The forces between kinks fall off exponentially fast and hence the dominant forces are between nearest neighbors. As discussed in the previous section, the sign of the force between the ith soliton (kink or antikink) and the $(i + 1)$th soliton (antikink or kink) is proportional to $\text{Tr}(Q_i Q_{i+1})$ where Q_i is the charge of the ith object. For the sequence above, $\text{Tr}(Q_i Q_{i+1}) > 0$ for every i and neighboring solitons repel each other. In particular, they cannot overlap and annihilate.

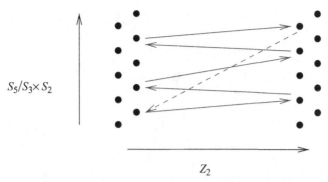

Figure 2.5   In the lattice of kinks of Eq. (2.36), the vacua are arranged sequentially in a pattern so as to return to the starting vacuum only after several transitions between the two discrete (Z_2) sectors.

The sequence of kinks in Eq. (2.35) has a period of six kinks. These six kinks have a net topological charge that vanishes since the last vacuum expectation value in Eq. (2.36) is the same as the first value. Hence we can put the sequence in a periodic box, i.e. compact space. This gives us a finite lattice of kinks.

The sequence described above has the minimum possible period (namely, six). It is easy to construct other sequences with greater periodicity. For example

$$\ldots Q^{(1)} \bar{Q}^{(5)} Q^{(3)} \bar{Q}^{(4)} Q^{(2)} \bar{Q}^{(1)} Q^{(5)} \bar{Q}^{(3)} Q^{(4)} \bar{Q}^{(2)} \ldots \tag{2.37}$$

is a repeating sequence of ten kinks.

The lattice of kinks is a solution in both the $S_5 \times Z_2$ and the $SU(5) \times Z_2$ models. However, it is stable in the former and unstable in the latter. The instability in the $SU(5) \times Z_2$ model occurs because a kink of a given charge, say $Q^{(3)}$, can change with no energy cost into a kink of some other charge, for example $Q^{(1)}$. Then, in the sequence of Eq. (2.35), the third kink changes into $Q^{(1)}$, then annihilates with the antikink with charge $\bar{Q}^{(1)}$ on its right. In this way the lattice can relax into the vacuum. In the $S_5 \times Z_2$ case, however, the degree of freedom that can change the charge of a kink is absent and the lattice is stable.

So far we have been discussing a kink lattice in one periodic dimension. This is equivalent to having a kink lattice in a circular space. Next consider what happens in a plane in two spatial dimensions. A circle in this plane can once again have a kink lattice since neighboring kinks and antikinks repel. However, when extended to the whole plane, the kink lattice must have a nodal point as shown in Fig. 2.6. In three spatial dimensions, the nodal points must extend into nodal curves.[5]

We shall discuss kink lattices further in Chapter 6.

---

[5] This is very similar to the case where several domain walls terminate on topological strings, except that there are no topological strings in the model.

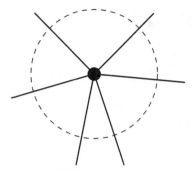

Figure 2.6   A domain wall lattice consisting of six domain walls can be formed in a one-dimensional sub-space (dashed circle) of a two-dimensional plane. This domain wall lattice is stable. Extending it to the two-dimensional plane, the different domain walls converge to a nodal point. This implies that the $S_5 \times Z_2$ model contains domain wall nodes (or junctions) in two dimensions and nodal curves in three spatial dimensions.

2.9 Open questions

1. Discuss all topological and non-topological kink solutions in an $SU(N) \times Z_2$ model where N is even. In [163] the case with odd N is discussed.[6]

---

[6] However, it is incorrectly stated that the Z_2 symmetry is included in $SU(N)$ when N is even, as can be seen from the $\text{Tr}(\Phi^3)$ argument of Section 2.1.

3

Interactions

In the previous two chapters, we have described kink solutions in several models but these solutions have mostly been discussed in isolation. In any real system, there is a variety of kinks and antikinks, in addition to small excitations (particles) of the fields. The interactions of kinks with other kinks and with particles play an important role in the evolution of the system. The motion of kinks is also accompanied by the radiation of particles. Ambient particles in the system scatter off kinks, and kinks collide with each other, and perhaps annihilate into particles. As discussed in Section 1.9, in some models a kink-antikink pair can bind together to form a non-dissipative solution which is called a "breather." In other models, approximate breather solutions have been found, which play an important role in the scattering of a kink and an antikink. These topics are discussed in the following sections.

3.1 Breathers and oscillons

So far we have been considering kinks, which are static solutions to the equations of motion. In the sine-Gordon model of Eq. (1.51), a one-parameter family of non-static, dissipationless solutions is also known. These are bound states of a kink and an antikink and are called breathers. The breather solution was described briefly in Section 1.9 and can be re written as

$$\phi_{\mathrm{b}}(t, x; v) = \frac{4}{\beta} \tan^{-1}\left[\frac{\sin(v\sqrt{\alpha}t/\sqrt{1+v^2})}{v\cosh(\sqrt{\alpha}x/\sqrt{1+v^2})} \right] \tag{3.1}$$

where v is a free parameter (see Fig. 3.1). We will have more to say about breathers when we quantize kinks in Chapter 4 as they play a very fundamental role in the novel duality between the sine-Gordon model and the massive Thirring model (see Section 4.7).

Breather solutions are not known to exist in the $\lambda\phi^4$ model [135]. However, numerical studies of the scattering of a Z_2 kink and antikink revealed the existence of

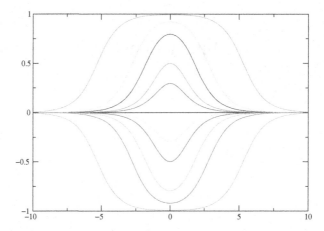

Figure 3.1 The sine-Gordon breather shown at various times during one oscil-lation period. At certain times, the field profile is that of a separated kink and an antikink. At other times, the kink and the antikink overlap and cannot be distin-guished.

extremely long-lived, oscillating bound states of kinks and antikinks [4, 19, 26, 64]. The existence of kink-antikink bound states has been interpreted as a resonance phenomenon between the natural excitation frequency of the kink profile (shape mode) and the frequency of oscillation of the bound kink-antikink system. Radiation from a time-dependent scalar field configuration will be suppressed if the oscillation frequency of the configuration is small compared to the mass of the radiation quanta and this can be used to understand the longevity of oscillons (Farhi, 2005, private communication).

The simplest hypothesis is that oscillons are approximate breather solutions since a region of the sine-Gordon potential and the $\lambda\phi^4$ potential have very similar shapes. We can compare the two potentials when the sine-Gordon potential has been shifted so that it has a maximum at $\phi = 0$. The parameter β in the sine-Gordon model is chosen so that the first positive minimum is at $\phi = +\eta$. α is fixed by requiring that the masses of small excitations in the true vacua, given by the second derivative of the potential, are equal in the two models. Then the two potentials are given by

$$V_{Z_2}(\phi) = \frac{\lambda}{4}(\phi^2 - \eta^2)^2 \tag{3.2}$$

$$V_{\text{sG}}(\Phi) = \frac{\alpha}{\beta^2}(1 - \cos(\beta(\phi - \eta))) \tag{3.3}$$

with

$$\alpha = 2\lambda\eta^2, \qquad \beta = \frac{\pi}{\eta} \tag{3.4}$$

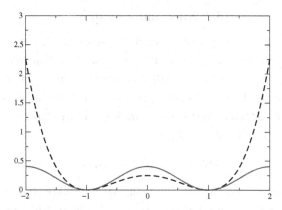

Figure 3.2   The $\lambda\phi^4$ potential (broken curve) and the shifted sine-Gordon potential (solid curve) when the parameters are chosen so that the vacua occur at the same values of ϕ and the curvatures of the potentials at the vacua are also equal.

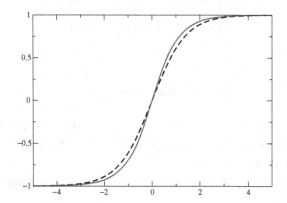

Figure 3.3   The profiles of the kinks in the $\lambda\phi^4$ model (broken curve) and the shifted sine-Gordon model (solid curve) with potentials as shown in Fig. 3.2.

The two potentials can be compared in the vicinity of their true vacuum at $\phi = \eta$. Then

$$V_{Z_2}(\phi) = \lambda\eta^2(\phi - \eta)^2 + \lambda\eta(\phi - \eta)^3 + \frac{\lambda}{4}(\phi - \eta)^4 \tag{3.5}$$

and

$$V_{sG}(\phi) = \lambda\eta^2(\phi - \eta)^2 - \frac{2\pi^2\lambda}{4!}(\phi - \eta)^4 + O((\phi - \eta)^6) \tag{3.6}$$

In Fig. 3.2 we show these two potentials and in Fig. 3.3 we compare the kink profiles.

   We will return to the breather and its role in the quantum sine-Gordon model at the end of Section 4.7.

3.2 Kinks and radiation

By "radiation" we mean propagating excitations of small amplitude of a field, which in this chapter will be taken to be the same field that makes up the kink. Asymptotically, these excitations have the usual plane wave form: $\exp(i(\omega t \pm kx))$. In the kink background, these "scattering states" are found as solutions to the equation of motion for fluctuations about the kink. If we denote the kink solution by $\phi_k(x)$, the fluctuation field $\psi(t, x)$ is

$$\psi(t, x) = \phi(t, x) - \phi_k(x) \tag{3.7}$$

We will assume $|\psi| \ll \langle\phi\rangle$, where $\langle\phi\rangle$ is the vacuum expectation value of ϕ. To find the scattering states, we take $\psi = f(x)e^{-i\omega t}$ where it is understood that the real or imaginary part should be taken – in other words, the physical modes are $[f(x)e^{-i\omega t} \pm f^*(x)e^{+i\omega t}]$. Perturbing the Lagrangian for ϕ (first line of Eq. (1.2)), we find that $f(x)$ satisfies the (linearized) equation of motion

$$Hf \equiv -f'' + U(x)f = \omega^2 f \tag{3.8}$$

where

$$U(x) \equiv V''(\phi_k(x)) \equiv \left.\frac{\partial^2 V}{\partial\phi^2}\right|_{\phi=\phi_k} \tag{3.9}$$

The scattering states around a static kink are obtained by solving the Schrödinger-type equation, Eq. (3.8), which for some potentials, falls in the general class of equations discussed in Appendix C.

We now consider the Z_2 kink for which the potential U is obtained from Eqs. (3.9) and (1.2) to be

$$U(x) = \lambda\left(3\phi_k^2 - \eta^2\right) \tag{3.10}$$

We now list the eigenmodes of Eq. (3.8). (We will encounter them again in Chapter 4.) First, there are two bound states, also known as "discrete" modes:

$$\omega_0 = 0, \qquad f_0 = \text{sech}^2 z \tag{3.11}$$

$$\omega_1 = \frac{\sqrt{3}}{2}m_\psi, \qquad f_1 = \sinh z \, \text{sech}^2 z \tag{3.12}$$

where $z = x/w = m_\psi x/2$. The $\omega = 0$ mode is called the "translation mode" and the second is the "shape mode." Then there is a continuum of states for $m_\psi < \omega < \infty$ which are the scattering states:

$$f_k = e^{ikx}[3\tanh^2 z - 1 - w^2 k^2 - i\,3wk\tanh z] \tag{3.13}$$

The dispersion relation is

$$\omega_k^2 = k^2 + m_\psi^2 \tag{3.14}$$

We are now interested in processes that involve both a kink and the scattering states (radiation). For example, if a kink accelerates, it will emit radiation. What is the radiated power? The answer will depend on the forces that make the kink accelerate and whether or not these forces deform the structure of the kink.[1] We shall examine the radiation from kink shape deformations and other interactions of kinks and radiation after a brief diversion in the next section.

3.3 Structure of the fluctuation Hamiltonian

In this section we will show two interesting properties of the fluctuation Hamiltonian, H, defined in Eq. (3.8). The first is that the potential $U(x)$ has a very special form that implies that the Hamiltonian can be factored. The second is that there exists a "partner Hamiltonian" with (almost) the same spectrum as the original Hamiltonian.

The special form of U follows from the fact that the kink has a translation zero mode (see Section 1.1). Hence there exists an eigenstate with $\omega = 0$. Denote this "translation mode" by ψ_t. Hence

$$H\psi_t = (-\partial^2 + U(x))\psi_t = 0 \tag{3.15}$$

Therefore

$$U(x) = \frac{\psi_t''}{\psi_t} \tag{3.16}$$

which can also be rewritten as

$$U(x) = f' + f^2, \quad f = (\ln(\psi_t))' \tag{3.17}$$

For the particular cases of the Z_2 and sine-Gordon kinks, not only is $U(x)$ of the form in Eq. (3.17) but it is also reflectionless. Then, an incident wave is fully transmitted and the reflection coefficient vanishes. In this case, the only non-trivial characteristic of scattering states is that the waves get a phase shift owing to the presence of $U(x)$. This property will be useful when we quantize the kink in Section 4.1.

The Hamiltonian H with a potential of the form in Eq. (3.17) has the important property that it can be factored

$$H = A^+A \equiv (+\partial + f)(-\partial + f) \tag{3.18}$$

Therefore the equation for the eigenstates is simply

$$Hf = A^+Af = \omega^2 f \tag{3.19}$$

---

[1] In the case of domain walls in three spatial dimensions, the curvature of the wall is itself responsible for acceleration. This motion leads to the emission of scalar and gravitational radiation and will be discussed in Chapter 8.

The factorization has the consequence that one can readily construct a "partner" Hamiltonian, H_-, that has almost an identical eigenspectrum as H. This partner Hamiltonian is $H_- = AA^+$. If f_i is an eigenstate of H with eigenvalue ω_i^2, then Af_i is an eigenstate of H_- with the same eigenvalue. This argument works for all eigenstates except the one for which $Af_i = 0$. Hence H has a single extra eigenstate with $\omega = 0$.[2]

The potential $U(x)$ determines the spectrum of excitations around a soliton. The factorizability of the Hamiltonian is useful in the problem of reconstructing $V(\phi)$ from the spectrum of fluctuations (i.e. the set of ω^2) using inverse scattering methods [165].

3.4 Interaction of kinks and radiation

As remarked below Eq. (3.17) the potentials $U(x)$ for both the Z_2 and the sine-Gordon kinks are rather special since they are reflectionless. All that happens is that the transmitted wave gets phase shifted. This is equivalent to a time delay in the propagation of the wave through the kink.

From the solution for the scattering states given in Eq. (3.13) for the Z_2 kink we find a momentum dependent phase shift

$$\delta_k|_{Z_2} = 2\tan^{-1}\left(\frac{3wk}{w^2k^2 - 2}\right) \tag{3.22}$$

This corresponds to a time delay

$$\tau_k\Big|_{Z_2} = \frac{\delta_k}{\omega}\Big|_{Z_2} = \frac{2}{\sqrt{k^2 + m_\psi^2}}\tan^{-1}\left(\frac{3wk}{w^2k^2 - 2}\right) \tag{3.23}$$

Similarly the phase shift and time delay in the case of the sine-Gordon kink are

$$\delta_k|_{sG} = \pi - 2\tan^{-1}\left(\frac{k}{m_\psi}\right) \tag{3.24}$$

$$\tau_k\Big|_{sG} = \frac{\delta_k}{\omega}\Big|_{sG} = \frac{1}{\sqrt{k^2 + m_\psi^2}}\left[\pi - \tan^{-1}\left(\frac{k}{m_\psi}\right)\right] \tag{3.25}$$

---

[2] The two partner Hamiltonians can also be combined to form a supersymmetric Hamiltonian, H_{ss}

$$H_{ss} = \begin{pmatrix} A^+A & 0 \\ 0 & AA^+ \end{pmatrix} = \{Q, Q^+\} \equiv QQ^+ + Q^+Q \tag{3.20}$$

where

$$Q = \begin{pmatrix} 0 & 0 \\ A & 0 \end{pmatrix}, \qquad Q^+ = \begin{pmatrix} 0 & A^+ \\ 0 & 0 \end{pmatrix} \tag{3.21}$$

While there is no reflection of radiation of the same field that makes up the kink in the Z_2 and sine-Gordon cases, there can be reflection of fluctuations of other fields [53]. As an example [171], consider a second scalar field χ included in the Z_2 model so that the full Lagrangian becomes

$$L = L_\phi + \frac{1}{2}(\partial_\mu \chi)^2 - \frac{m_\chi^2}{2}\chi^2 - \frac{\sigma}{2}\phi^2\chi^2 \tag{3.26}$$

where L_ϕ is the Lagrangian for the Z_2 model (Eq. (1.2)). Then the scattering modes of χ in the presence of a Z_2 kink are found by solving

$$\partial_t^2\chi - \partial_x^2\chi + m_\chi^2\chi + \sigma\phi_k^2\chi = 0 \tag{3.27}$$

Substituting $\phi_k = \eta\tanh(x/w)$ and $\chi = \exp(-i\omega t)f(x)$, we get

$$\partial_X^2 f + (v^2 - \bar{\sigma}\,\mathrm{sech}^2(X))f = 0 \tag{3.28}$$

where $X \equiv x/w$, $v^2 = w^2(\omega^2 - m_\chi^2 - \sigma\eta^2)$, $\bar{\sigma} = \sigma\eta^2 w^2$. (Recall that $w = \sqrt{2/\lambda\eta^2}$.)

Equation (3.28) is a special case of the differential equation described in Appendix C. The scattering state is found for real values of v and has the asymptotics: $f \to e^{ikx}$ for $x \to \infty$, and for $x \to -\infty$:

$$f \to \frac{\Gamma(1-ikw)\Gamma(-ik)e^{ikx}}{\Gamma(1/2+\gamma-ik)\Gamma(1/2-\gamma-ik)} + \frac{\Gamma(1-ikw)\Gamma(ik)e^{-ikx}}{\Gamma(1/2+\gamma)\Gamma(1/2-\gamma)} \tag{3.29}$$

where $k = v/w$ and $\gamma = \sqrt{\bar{\sigma}+1/4}$.

The reflection coefficient can be read off from the asymptotic behavior of $f(x)$ as $x \to -\infty$ and has been evaluated in Section 12.3 of [113]

$$R = \frac{1+\cos(2\pi\gamma)}{\cosh(2\pi k) + \cos(2\pi\gamma)} \tag{3.30}$$

The transmission coefficient is

$$T = \frac{2\sinh^2(\pi k)}{\cosh(2\pi k) + \cos(2\pi\gamma)} = 1 - R \tag{3.31}$$

From the asymptotic expression in Eq. (3.29), it is also possible to calculate the time delay of the reflected and transmitted waves owing to the kink. For example, if we write

$$\frac{\Gamma(1-ikw)\Gamma(-ik)}{\Gamma(1/2+\gamma-ik)\Gamma(1/2-\gamma-ik)} = |T|^{1/2}e^{i\delta_k} \tag{3.32}$$

where T is the transmission coefficient above, then the time delay of the transmitted wave is given by δ_k/ω.

3.5 Radiation from kink deformations

A static kink does not emit any radiation. Nor does it emit radiation if it is moving at constant velocity (see Eq. (1.10)). However, if the kink is accelerating (owing to some external force), or its shape is deformed, it can emit radiation in the form of scalar particles [106, 107]. In 3 + 1 dimensions, acceleration and deformations arise since the kinks (domain walls) are moving under their own tension, except in the very special cases of static solutions. The radiation emitted from curved domain walls has not been calculated analytically, though the problem has been studied numerically [182]. In the case of 1 + 1 dimensional Z_2 kinks that are undergoing periodic deformations, the radiation has been found analytically in [110, 140], and we shall describe this calculation below.

Following [110], we simplify notation by setting $\lambda = 2$ and $\eta = 1$ in the Z_2 model so that $w = 1$ in these units (see Section 1.1). Then the field $\phi(x, t)$ is written in terms of the complete set of small excitations. This gives

$$\phi(x, t) = \phi_k(x) + R(t) f_0(x) + A(t) f_1(x) + f(x, t) \tag{3.33}$$

where $\phi_k = \tanh(x)$, f_0 and f_1 are the translation and shape modes respectively as given in Eqs. (3.11) and (3.12), $R(t)$ and $A(t)$ are their time-dependent amplitudes, and the function $f(x, t)$ contains all the continuum states around the kink. The frequency of oscillation of $R(t)$ is $\omega_0 = 0$ and of $A(t)$ is $\omega_1 = \sqrt{3}$. These values were derived for linearized fluctuations about the kink. Non-linearities will modify $\omega_1 = \sqrt{3}$ but we assume that such modifications are small.

We will work in the rest frame of the kink and so

$$R(t) = 0 \tag{3.34}$$

The idea now is to insert Eq. (3.33) in the equation of motion for ϕ with some choice of the amplitude $A(t)$ which is assumed to be small, and then to find the solution for the scattering states, $f(x, t)$, which form the radiation.

Insertion of Eq. (3.33) in Eq. (1.4) gives

$$(\ddot{A} + 3A) f_1 + \ddot{f} - f'' + 2(3\phi_k^2 - 1) f = -6(f + \phi_k) f_1^2 A^2$$
$$- 6(f + 2\phi_k) f f_1 A - 2 f_1^3 A^3$$
$$- 6\phi_k f^2 - 2 f^3 \tag{3.35}$$

where the equations satisfied by ϕ_k and f_1 have been used. Assuming that A is small, and that f is $O(A^2)$ or smaller, the leading order equation is $\ddot{A} + 3A = 0 + O(A^2)$. Then to order A^2, the equation for f is

$$(\ddot{A} + 3A) f_1 + \ddot{f} - f'' + 2(3\phi_k^2 - 1) f = -6\phi_k f_1^2 A^2 \tag{3.36}$$

The f-independent terms on the right-hand side of Eq. (3.35) are source terms which cause radiation. Hence f will not be zero at order A^2. The terms will also cause the amplitude, A, of the shape mode to depart from the purely oscillatory behavior. To determine how much of the source affects radiation and how much affects the shape mode, note that f_1 and f are orthogonal

$$\int dx f_1(x) f(x, t) = 0 \tag{3.37}$$

So we can decompose the equation into a direction parallel to f_1 in mode space and orthogonal to it. One assumption we have to make is that the back-reaction of the radiative modes on the shape mode is higher order in A. For example, Eq. (3.37) does not by itself imply that f'' and f_1 are orthogonal. Then, multiplying Eq. (3.36) by f_1 and integrating over all space gives

$$\ddot{A} + 3A = -6A^2 \int dx \phi_k f_1^2 \equiv -6\alpha A^2 \tag{3.38}$$

provided we have normalized f_1 so that

$$\int dx [f_1(x)]^2 = 1 \tag{3.39}$$

Explicit evaluation gives

$$\alpha = \frac{3\pi}{32} \tag{3.40}$$

The equation orthogonal to f_1 is

$$\ddot{f} - f'' + 2(3\phi_k^2 - 1)f = -6\phi_k f_1^2 A^2 + 6\alpha f_1 A^2 \tag{3.41}$$

and this will determine the radiation from the deformed kink once we have specified A.

The leading order solution for A is

$$A = A_0 \cos(\sqrt{3}t) \tag{3.42}$$

Hence

$$A^2 = \frac{A_0^2}{2}[\cos(2\sqrt{3}t) + 1] \tag{3.43}$$

This form implies that the source for f in Eq. (3.41) has a time-dependent piece and another time-independent piece. Since the equation is linear in f, only the time-dependent piece proportional to $\cos(2\sqrt{3}t)$ is important. Setting

$$f(x, t) = \text{Re}(e^{i\omega t} F(x)) \tag{3.44}$$

the equation that needs to be solved is

$$-F'' + \left(6\phi_k^2 - 2 - \omega^2\right)F = \frac{3}{2}\left(\alpha f_1 - \phi_k f_1^2\right)A_0^2 e^{i(\omega_0 - \omega)t} \tag{3.45}$$

where $\omega_0 = 2\sqrt{3}$. Since the left-hand side is time-independent, this only has solutions for

$$\omega = \omega_0 = 2\sqrt{3} \tag{3.46}$$

and then all the solutions of the homogeneous equation are known (see Appendix C; [113, 126]). The solutions of the homogeneous equation with plane wave asymptotics are

$$F_q(x) = \left(3\phi_k^2 - 1 - q^2 - 3iq\phi_k\right)e^{iqx} \tag{3.47}$$

where $q = \sqrt{\omega^2 - 4}$. Knowing all the solutions of the homogeneous equation, it is possible to explicitly construct the (retarded) Green's function suitable for outgoing radiation.

$$G(x, y) = \begin{cases} -F_{-q}(y)F_q(x)/W, & (x < y) \\ -F_q(y)F_{-q}(x)/W, & (x > y) \end{cases} \tag{3.48}$$

where W is the Wronskian

$$W = F_q(x)F_{-q}'(x) - F_q'(x)F_{-q}(x) \tag{3.49}$$

The Wronskian is a constant and its value can be found by using the explicit solutions

$$W = -2iq(q^2 + 1)(q^2 + 4) \tag{3.50}$$

The solution of the inhomogeneous equation (3.45) is found by convoluting the source with the Green's function

$$F(x) = \int_{-\infty}^{+\infty} dy\, G(x, y)\frac{3}{2}\left[\alpha f_1(y) - \phi_k(y)f_1^2(y)\right]A_0^2 \tag{3.51}$$

With a little more manipulation, we obtain the radiation field in the $x \to +\infty$ limit

$$f(x, t) = \text{Re}\left[\frac{-3A_0^2 e^{i(\omega t - qx)}}{2iq(2 - q^2 - 3iq)}\int_{-\infty}^{+\infty} dy\, \phi_k(y)f_1^2(y)F_q(y)\right] \tag{3.52}$$

with $\omega = \omega_0 = 2\sqrt{3}$ and $q = \sqrt{\omega^2 - 4} = 2\sqrt{2}$. The integral can be done explicitly leading to

$$f(x, t) = \frac{\pi q(q^2 - 2)}{32\sinh(\pi q/2)}\sqrt{\frac{q^2 + 4}{q^2 + 1}}A_0^2\cos(\omega t - qx - \delta) \tag{3.53}$$

The phase δ can be read off from Eq. (3.52) because the integral is purely imaginary and does not contribute

$$\delta = \tan^{-1}\left(\frac{3q}{q^2 - 2}\right) \tag{3.54}$$

Now that we have the solution for the radiation field, we can find the energy flux by using the T_{0i} components of the energy-momentum tensor in Eq. (1.39). Including a factor of 2 to account for the radiation toward $x \to -\infty$, we obtain the radiated power [110]

$$\frac{\mathrm{d}E}{\mathrm{d}t} = -0.020A_0^4 \tag{3.55}$$

The back-reaction of the radiation on the deformation amplitude can be estimated on the grounds of energy conservation. In [110] the results above are compared to the results of a numerical evolution of the deformations using the full non-linear equations with good agreement.

3.6 Kinks from radiation

By time reversing kink and antikink annihilation, it should be possible to obtain kink-antikink creation from incoming radiation. However, the stream of incoming radiation would have to be sent with just the correct phase relationship and energy. Such initial conditions occupy zero volume in the space of all initial conditions. A more physical problem is to identify the set, or a large subset, of initial conditions for the incoming radiation that will lead to kink-antikink creation. This problem is unsolved. Yet certain interesting results have been obtained in [110] in the "gradient flow" approximation in which the second time derivative terms in the equation of motion are neglected.

Consider the collision of two kinks in the presence of a pre-existing kink [110], as depicted in Fig. 3.4. As the kink-antikink-kink ($k\bar{k}k$) system evolves, a kink-antikink annihilate, and we are left with a kink whose shape is excited. Reversing this process, if we start with a kink whose shape is excited, it can produce a kink-antikink pair. In [110], this relation between the shape mode and the creation of a kink-antikink pair was explored.

3.7 Scattering of kinks

The sine-Gordon model is a famous example of a completely integrable system [48]. Sine-Gordon kinks are examples of "solitons" in the strict mathematical sense in which when two or more solitons (or anti-solitons) scatter, they simply pass through

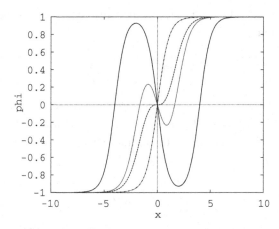

Figure 3.4 A kink collides from the left with another kink coming in from the right in the presence of an antikink in the middle. The time evolution of the field is shown in succession by the solid, dotted, dashed, and dashed-dotted curves. The evolution shows that a kink and antikink annihilate leaving behind a kink whose shape mode is excited (dotted and dashed curves). With further evolution, the shape mode will dissipate and an unexcited kink will remain as seen in the dashed-dotted curve. [Figure reprinted from [110].]

each other. The only consequence of the scattering is that there is a phase shift, or equivalently, a time delay. The time delay may be understood by realizing that the force between two kinks in the sine-Gordon model is attractive. Hence the kinks collide and form a bound state for some time. The time delay may be viewed as the time spent by the kinks in the form of a bound state. A crucial aspect of the scattering is that there is no dissipation. More details can be found, for example, in [48].

Kink scattering in the Z_2 model has a more complex character. In this case, we cannot have kink-kink scattering because two Z_2 kinks cannot be adjacent to each other. Instead, we need only consider kink-antikink scattering. This has been the subject of significant investigation [26, 4]. When a kink-antikink collide, the only possibilities are that they reflect back or they annihilate (see Fig. 3.5).

As we might expect, at very low incoming velocities, a bound state is formed and annihilation inevitably occurs, while at very high velocities, reflection takes place. The remarkable discovery of numerical studies of kink-antikink scattering is that the change from annihilation to reflection does not happen at just one critical value of the incoming velocity. Instead there are bands of incoming velocity at which annihilation occurs, while at other values of the incoming velocity the kink and antikink are reflected. The plot in Fig. 3.6 shows these results.

The unexpected dependence of kink-antikink scattering on the incoming velocity has been examined closely in [26, 4]. The behavior is understood as a resonance effect between oscillations of the mode that describes the shape distortions of the

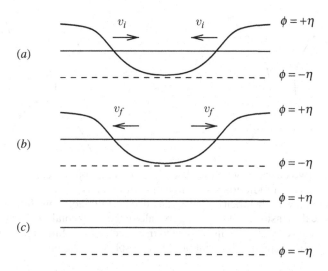

Figure 3.5 A kink and an antikink with incoming velocity v_i are shown in panel (*a*). The two possible outcomes of the scattering are shown in panels (*b*) and (*c*). In panel (*b*), the kinks scatter and reflect. Their outgoing velocity v_f need not be equal to v_i. In panel (*c*), the kink and antikink have annihilated and radiated away their energy, leaving behind the trivial vacuum. In both outcomes, the scattering is likely to be accompanied with radiation that has not been depicted.

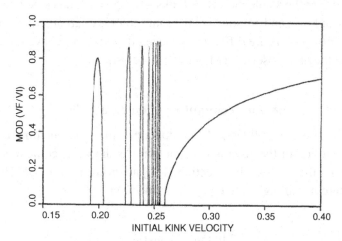

Figure 3.6 The ratio of outgoing to incoming kink velocities after scattering versus the incoming velocity [26, 4]. When the outgoing velocity is plotted to be zero, a kink-antikink bound state is formed that decays to the vacuum by radiation. Notice that the kink-antikinks annihilate in certain bands in the initial velocity. [Figure reprinted from [26].]

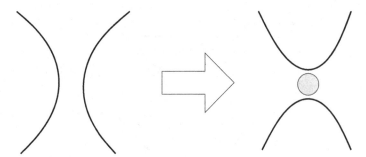

Figure 3.7 Two curved domain walls collide and intercommute. At the collision point, there is lots of radiation owing to annihilation or owing to the formation of a closed domain wall that then collapses and decays into radiation. To imagine the walls in three dimensions, rotate the figures along the horizontal axis. In the initial state the two curved walls are disconnected from each other while in the final state, the wall is in the shape of a "wormhole," with a sphere in the middle.

kinks and the oscillations of the kinks as a whole owing to kink-antikink interactions. We shall not describe the details of the analysis here.

The scattering of $SU(5) \times Z_2$ kinks has been studied numerically in [121]. In this case, there is an additional degree of freedom, namely the non-Abelian charge of the colliding kinks (or "color") and there are a variety of initial conditions that can be considered. For the stable variety of kinks – the $q = 2$ kinks (see Section 3.2) – the scattering of kinks and antikinks of the same color is qualitatively similar to that of Z_2 kinks. If the colors are different, however, there is a repulsive force between the kinks and they are observed to bounce back elastically.

3.8 Intercommuting of domain walls

We finally consider the collision of two domain walls. The outcome is found by numerical evolution of the equations of motion. As the walls come together, they reconnect along the curve of intersection [136] as shown in Fig. 3.7. This process is called "intercommuting" or, simply, "reconnection."

3.9 Open questions

1. Suppose we want to create a well-separated Z_2 kink-antikink pair by colliding small amplitude plane waves (particles) in the $\phi = +\eta$ vacuum. What conditions must be imposed on the incoming waves? What is the space of initial conditions that leads to domain wall formation? Can the initial conditions be implemented in a practical setting (e.g. accelerator experiments)?
2. Can a domain wall lattice be generalized to other defects, e.g. a lattice of strings and monopoles?

3. Study the interaction of domain walls and strings/magnetic monopoles in a model that contains both types of defects e.g. the $SU(5) \times Z_2$ model has walls and magnetic monopoles.

4. Construct a sine-Gordon-like breather field configuration in the $\lambda\phi^4$ model. This will not be an exact solution of the field equation of motion. Hence it will radiate. Calculate the radiated power. Are there circumstances in which the radiated power is very small?

5. Can the analysis of radiation from kink deformations be extended to the case of oscillating domain walls? The simplest procedure would be to decompose the field as in Eq. (3.33) and to include a suitable external (harmonic) potential that drives the translation mode only. This will cause the kink to oscillate as a whole without deformations. However, the oscillations will source the shape mode and the radiation, and an analysis of the kind in Section 3.5 seems feasible.

6. Can the analysis of radiation from kink deformations be extended to the case of spherical domain walls?

7. How can the radiation analysis be extended to vortex solutions in two or more spatial dimensions?

4

Kinks in quantum field theory

A particle in a classical harmonic oscillator potential, $m\omega^2 x^2/2$, has minimum energy when it sits at rest at the bottom of the potential. Then the particle's energy vanishes. The Heisenberg uncertainty principle however modifies this picture for the quantum harmonic oscillator. The particle cannot sit at rest (with definite momentum) at the bottom of the potential (a definite location). Indeed, the quantum zero point motion lifts the ground state energy to $\omega/2$. Further, the excited states of the simple harmonic oscillator are discrete and occur at energies $(n + 1/2)\omega$, $n = 0, 1, 2, \ldots$

Just as the classical harmonic oscillator is modified by quantum effects, any classical solution to a field theory is also modified by quantum effects. Quantum effects give corrections to the classical kink energy owing to zero point quantum field fluctuations. These quantum corrections are small provided the coupling constant in the model is weak. To "quantize the kink" means to evaluate all the energy levels of the kink (first quantization) and to develop a framework for doing quantum field theory in a kink background. This involves identifying all excitations in the presence of the kink and their interactions. The field theory of the excitations in the non-trivial background of the kink is akin to second quantization. Finally, one would also like to describe the creation and annihilation of kinks themselves by suitable kink creation and annihilation operators. This would be the elusive third quantization.

Initially we calculate the leading order quantum corrections to the energy of the Z_2 and sine-Gordon kinks. As these two examples illustrate, the precise value of the quantum correction depends on the exact model and kink under consideration. Yet there is one common feature – quantum corrections tend to reduce the energy of the kink. This result is quite general and we prove it using a variational argument in Section 4.5.

The quantum corrections to the kink mass are obtained by using a perturbative analysis where the coupling constant is the expansion parameter, as first done in [38, 42]. For fixed values of the masses of particles in the field theory, the energy of the classical solution is proportional to the coupling constant raised to a negative power (for example, see Eq. (1.20)) and so the perturbative analysis holds only if the kink is much more massive than the particles in the model. As the coupling constant is increased, quantum effects become stronger and eventually the perturbative scheme breaks down. Remarkably, the sine-Gordon model is still amenable to analysis in this regime and, at strong coupling, the sine-Gordon kinks become lighter than the particles. Indeed, there exists a weakly coupled description of the model in which perturbative methods can be used: this is the massive Thirring model in which the particles (low energy excitations of a fermionic field) correspond to the sine-Gordon kinks at strong coupling (see Section 4.7).

The phenomenon observed in the sine-Gordon and massive Thirring models, in which solitons of one model (Model 1) are identified with the particles of a second model (Model 2) and vice versa in certain regimes of the coupling constants, is known as "duality." Model 1 is said to be dual to Model 2 if the particle-plus-soliton spectrum of Model 1 maps onto the soliton-plus-particle spectrum of Model 2 and vice versa. Both models describe the same physics, except that the light and heavy degrees of freedom are interchanged.

The Z_2 model does not share the remarkable symmetries of the sine-Gordon model and less is known about the Z_2 kink at strong coupling. However, the mass of the Z_2 kink can be evaluated at strong coupling using lattice field theory. We describe these results in Section 4.8 and conclude, once again, that the kink becomes less massive as the coupling is increased and eventually becomes massless.

In this book, we only describe quantization of the mass of the kink using canonical techniques. A more extensive discussion of various other techniques and issues can be found in [35, 126] and in the series of papers in [38, 42].

4.1 Quantization of kinks: broad outline

In this section, we evaluate the contribution of the zero point fluctuations to the energy of the kink. Then we briefly discuss excited states.

The quantization procedure can be outlined as follows:

- Consider a field theory in two dimensions with compact spatial dimension of size L, assumed large compared to any other length scale in the problem. Periodic boundary conditions are imposed on the fields. Eventually take $L \to \infty$.
- Consider small quantum fluctuations, ψ, about the classical kink background, ϕ_k,

$$\phi(t, x) = \phi_k(x) + \psi(t, x) \tag{4.1}$$

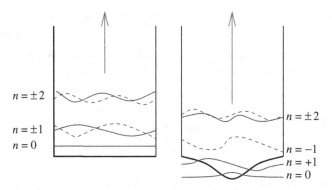

Figure 4.1   A trivial potential on a periodic space with period L is shown on the left. The field modes are labeled by an integer $n = 0, \pm1, \pm2, \ldots$ When there is a kink, the potential felt by the modes becomes non-trivial as depicted by the curved bottom of the figure on the right. What used to be the $n = 0$ mode in the trivial potential (on the left) becomes the lowest bound state, the zero mode, in the non-trivial potential. Similarly a linear combination of the $n = \pm1$ modes in the trivial box may become a second bound state ($n = +1$ in this illustration) and the other states remain unbound but shift in form and energy.

Linearize the equation for ψ and then quantize, that is, regard the field ψ as a quantum operator

$$\psi(t, x) = \sum_k [a_k f_k(t, x) + a_k^\dagger f_k^*(t, x)] \tag{4.2}$$

where $a_k^\dagger$ and a_k are creation and annihilation operators. The f_k are mode functions i.e. orthonormal solutions of the linearized equations of motion for ψ in the kink background. The equation satisfied by f_k is

$$\partial_t^2 f_k - \partial_x^2 f_k + V''(\phi_k(x)) f_k = 0 \tag{4.3}$$

- Find all the eigenmodes, f_k, of the fluctuations and their eigenfrequencies ω_k. As shown in Fig. 4.1, in the presence of the kink the modes are displaced. Some of the low-lying modes without the kink become bound states in the presence of the kink, and the others become scattering states as $L \to \infty$.
- Each eigenmode corresponds to a quantum harmonic oscillator with zero point fluctuations. Sum up the zero point energies of all the modes to get the quantum correction to the classical kink energy, E_{cl},

$$\tilde{E} = E_{cl} + \sum_i \frac{1}{2} \omega_i \tag{4.4}$$

In the $L \to \infty$ limit, the sum over the modes becomes a sum over bound states and an integral over scattering states. Also note that Eq. (4.4) is only valid to leading order in the quantum corrections since we have ignored interactions of the fluctuation field, ψ.

In following this procedure, quantum field theoretic subtleties arise.

- The zero point energy of the trivial vacuum (without the kink) must be subtracted from the zero point energy of the kink since we want to define the energy of the trivial vacuum to be zero. Therefore

$$E = E_{cl} + \sum_i \frac{1}{2}\omega_i - \left[E_{cl,0} + \sum_i \frac{1}{2}\omega_i^{(0)} \right]$$

where $E_{cl,0}$ is the classical energy of the trivial vacuum and is chosen to vanish ($E_{cl,0} = 0$), and $\omega_i^{(0)}$ are the eigenfrequencies of the modes in the trivial vacuum.
- The energy must be expressed in terms of renormalized parameters.

In the trivial vacuum, the energy eigenvalue for the mode with n nodes is

$$\omega_n^{(0)} = \sqrt{k_n^2 + m_\psi^2} \tag{4.5}$$

where $k_n = 2\pi n/L$ and $n \in \mathbf{Z}$, the set of all integers. Now suppose that the kink potential V'' is turned on slowly, i.e. that the potential term in Eq. (4.3) is multiplied by a parameter that vanishes for the free field theory and is continuously increased to one to get to the kink case. As the parameter increases, modes in the trivial box evolve into modes in the kink background. Some of the low-lying modes in the trivial box become the bound states of the kink. Let us label these modes by the index b (for "bound") and the remaining modes by c (for "continuum"). (In the example of Fig. 4.1, $b = 0, 1$, and c is any integer except for 0, 1.) Then

$$E = E_{cl} + \frac{1}{2}\sum_b \left(\omega_b - \omega_b^{(0)} \right) + \frac{1}{2}\sum_c \left[\sqrt{p_c^2 + m_\psi^2} - \sqrt{k_c^2 + m_\psi^2} \right] \tag{4.6}$$

where $\omega_c \equiv \sqrt{p_c^2 + m_\psi^2}$ and m_ψ denotes the mass of the ψ particles. In the limit $L \to \infty$, the sum over continuum states becomes an integral.

The terms in Eq. (4.6) can be understood quite simply. The first term on the right is the classical kink energy, the second contains the excess quantum corrections owing to the zero point motion of the modes bound to the kink, and the third term is the excess energy in the zero point motion of the modes that are not bound to the kink. The wave numbers of the scattering modes in the background of the kink are denoted by p_ν while that of the modes in the trivial vacuum by k_n.

In the trivial vacuum and when $L \to \infty$, the scattering states are plane waves, which are both energy and momentum eigenstates with $k_n = 2\pi n/L$. In the presence of the kink, the scattering states are energy eigenstates but not momentum eigenstates and, in general, an incoming wave gets both reflected and transmitted. Without specifying the field theory, further progress is possible when the scattering potential, $V''(\phi_k(x))$, is *reflectionless*. This may seem very restrictive, but it holds for both the Z_2 and sine-Gordon models and we assume it to be true for the remainder of this analysis. Then, asymptotically, the scattering states behave as

$$e^{i(px - \alpha(p)/2)} \text{ as } x = -L/2 \to -\infty \tag{4.7}$$

$$e^{i(px + \alpha(p)/2)} \text{ as } x = +L/2 \to +\infty \tag{4.8}$$

where $\alpha(p)$ is a phase shift. Note that on multiplying by $\exp(i\alpha(0)/2)$, the $p = 0$ state can be chosen to be purely real at $x = -\infty$. Since the scattering potential, $V''(\phi_k(x))$, is also real, this implies that the imaginary part of the wavefunction can be taken to be zero everywhere. Therefore $\alpha(0) = 0$.

The phase of the scattering states has a winding number given by the total phase change across the box. Since we have imposed periodic boundary conditions, the total phase winding, $(pL + \alpha(p))/2\pi$, must be an integer. This quantizes p so that

$$p_\nu L + \alpha(p_\nu) = 2\nu\pi \tag{4.9}$$

where $\nu \in \mathbf{Z}$ and we have denoted the νth wave-vector by p_ν.

Now that the scattering states in the soliton potential have been labeled by the integer ν, and those when the potential vanishes by the integer n, correspondence must be drawn between ν and n. To illustrate the problem, consider the $p_\nu = 0$ mode. As discussed above, $\alpha(0) = 0$ and hence, from Eq. (4.9), $\nu = 0$ labels this mode. Further, this mode has the lowest energy of the continuum states. In the specific example of Fig. 4.1, this mode corresponds to the $n = -1$ mode in the trivial box since the $n = 0, 1$ modes have become bound and have dropped out of the set of scattering states. Therefore $n = -1$ corresponds to $\nu = 0$, in this example.

With $k_n = 2\pi n/L$, we can write

$$p_\nu L + \Delta(p_\nu) = 2n\pi = k_n L \tag{4.10}$$

where

$$\Delta(p_\nu) \equiv \alpha(p_\nu) + 2\pi(n - \nu) \tag{4.11}$$

The shift in going from n to ν is the change in the total winding of the phase as the potential evolves from the trivial box to the soliton potential (see Fig. 4.1). As long as there is no change in the relative ordering of the energy levels, the heirarchy of the energy levels is maintained, and the mapping between n and ν is a constant shift. Since some of the low-lying states in the trivial potential have dropped out from the set of continuum states and have been converted into bound states, the set of integers n is partitioned into two subsets – one for the integers that lie above the would-be bound states and another for the integers that lie below the would-be bound states. The map from n in each subset to ν is a constant shift but the shift is different in the two subsets.

Next we think of $n - v$ as a function of k. In the $L \to \infty$ limit, $(n - v)$ is constant everywhere except at $k = 0$, meaning that the derivative of $n - v$ with respect to k is a Dirac delta function at $k = 0$,

$$\frac{\mathrm{d}\Delta}{\mathrm{d}k} = \text{coeff.}\delta(k) + \frac{\mathrm{d}\alpha(k)}{\mathrm{d}k} \qquad (4.12)$$

To determine the coefficient of the delta function, let us denote by N_b the number of states in the trivial potential that have dropped out as bound states in the kink potential. In a large interval $n_+ - n_-$ (n_+ is positive and n_- is negative), the corresponding interval in v is smaller by N_b, and hence the coefficient of the delta function is given by $-2\pi N_b$

$$\frac{\mathrm{d}\Delta}{\mathrm{d}k} = -N_b 2\pi \delta(k) + \frac{\mathrm{d}\alpha(k)}{\mathrm{d}k} \qquad (4.13)$$

For large momenta (and energy) the modes are unaffected by the deformation of the potential at the bottom of the well. Hence $p_v \to k_n$ in this region and $\Delta(p) \to 0$ as $|p| \to \infty$.

The phase shift $\Delta(p_v)$ depends on the potential in the equation of motion for $\psi(t, x)$, as in Eq. (4.3). As we explain below, the scattering potential created by the soliton background is non-perturbative. Therefore the phase shifts need not be small owing to factors of the coupling constant. However, note that $\Delta(p_v)/L$ is small as $L \to \infty$ and we need only keep terms up to linear order in $1/L$. Therefore

$$\sqrt{p_v^2 + m_\psi^2} = \sqrt{\left(k_n - \frac{\Delta(p_v)}{L}\right)^2 + m_\psi^2}$$

$$= \sqrt{k_n^2 + m_\psi^2} - \frac{k_n \Delta(k_n)}{L\sqrt{k_n^2 + m_\psi^2}} + O\left(\frac{1}{L^2}\right) \qquad (4.14)$$

Note that in the last line $\Delta(p_v)$ has been replaced by $\Delta(k_n)$ since $p_v = k_n + O(1/L)$.

We now want to express the energy of the kink in terms of renormalized parameters. If we denote the renormalized mass of ψ by $m_{\psi,\mathrm{R}}$ and the bare mass by $m_{\psi,\mathrm{b}}$, then

$$m_{\psi,\mathrm{R}}^2 = m_{\psi,\mathrm{b}}^2 - \delta m_\psi^2 \qquad (4.15)$$

where δm_ψ denotes the quantum contribution of vacuum fluctuations to the mass of ψ, and δm_ψ^2 is due to the self-coupling of the field and hence is proportional to the coupling constant.

The expression for the energy in Eq. (4.6) is valid to leading order in quantum corrections. The classical energy is inversely proportional to the coupling constant (e.g. Eq. (1.20)) and so the leading corrections are independent of the coupling constant. Note that m_ψ in the last two terms in Eq. (4.6) can be freely replaced by $m_{\psi,\mathrm{R}}$ since we are only evaluating the lowest order (coupling constant independent) quantum correction to the energy and δm_ψ^2 is proportional to the coupling constant. Retaining only the terms that are of leading order in the coupling constant, expanding E_{cl} in δm_ψ^2, and using Eq. (4.14), we get

$$E = E_{\mathrm{cl}}[m_{\psi,\mathrm{R}};\lambda_\mathrm{R}] + \Delta E_{\mathrm{cl}} + \frac{1}{2}\sum_b \left(\omega_b - \omega_b^{(0)}\right) - \frac{1}{2L}\sum_n \frac{k_n \Delta(k_n)}{\sqrt{k_n^2 + m_{\psi,\mathrm{R}}^2}} \qquad (4.16)$$

where ΔE_{cl} denotes the leading order change in E_{cl} when replacing bare parameters by renormalized parameters.

In the limit $L \to \infty$, the sum over n becomes an integral

$$\sum_n \to \frac{L}{2\pi}\int_{-\infty}^{+\infty} dk \qquad (4.17)$$

Hence,

$$E = E_{\mathrm{cl}}[m_{\psi,\mathrm{R}};\lambda_\mathrm{R}] + \Delta E_{\mathrm{cl}} + \frac{1}{2}\sum_b \omega_b - \frac{N_b}{2}m_{\psi,\mathrm{R}} - \frac{1}{4\pi}\int dk \frac{k\Delta(k)}{\sqrt{k^2 + m_{\psi,\mathrm{R}}^2}} \qquad (4.18)$$

where we have made use of the fact that $\omega_b^{(0)} = \sqrt{k_b^2 + m_{\psi,\mathrm{R}}^2} \to m_{\psi,\mathrm{R}}$ as $k_b \propto 1/L \to 0$.

On integration by parts

$$\int dk \frac{k\Delta(k)}{\sqrt{k^2 + m_{\psi,\mathrm{R}}^2}} = \left[\Delta(k)\sqrt{k^2 + m_{\psi,\mathrm{R}}^2}\right]_{-\infty}^{+\infty} - \int dk \sqrt{k^2 + m_{\psi,\mathrm{R}}^2}\,\frac{d\Delta}{dk} \qquad (4.19)$$

Since $\Delta(k)$ vanishes as $k \to \pm\infty$, the boundary term gives a finite contribution. The last term contains the derivative of $\Delta(k)$ and is given in Eq. (4.13). Therefore the final result is

$$E = E_{\mathrm{cl}}[m_{\psi,\mathrm{R}};\lambda_\mathrm{R}] + \Delta E_{\mathrm{cl}} + \frac{1}{2}\sum_b \omega_b - \frac{1}{4\pi}[|k|\Delta(k)]_{-\infty}^{+\infty}$$

$$+ \frac{1}{4\pi}\int dk \sqrt{k^2 + m_{\psi,\mathrm{R}}^2}\,\frac{d\alpha}{dk} \qquad (4.20)$$

Our general calculation can be pushed a little further since, in one spatial dimension, all divergences can be removed by normal ordering a "renormalized potential,"

V_R, which can be written in terms of the bare potential, $V(\phi)$ [35]

$$V_R = \exp\left\{\frac{1}{8\pi}\left(\ln\frac{4\Lambda^2}{m^2}\right)\frac{d^2}{d\phi^2}\right\}V(\phi) + \epsilon_0 \qquad (4.21)$$

where Λ is a momentum cut-off, and m is the bare mass. The constant ϵ_0 renormalizes the vacuum energy, and is chosen so that the expectation value of the Hamiltonian in the ground state vanishes. For example, in $\lambda\phi^4$ theory (Eq. (1.2)),

$$V_R = [\gamma(3\gamma\lambda - m^2) + \epsilon_0] + \frac{1}{2}(6\gamma\lambda - m^2)\phi^2 + \frac{\lambda}{4}\phi^4 \qquad (4.22)$$

where

$$\gamma \equiv \frac{1}{8\pi}\ln\left(\frac{4\Lambda^2}{m^2}\right) \stackrel{\Lambda\to\infty}{=} \frac{1}{8\pi}\int_{-\Lambda}^{+\Lambda}\frac{dk}{\sqrt{k^2 + m^2}} \qquad (4.23)$$

Then, the quantum correction to the mass is $\delta m^2 = 6\lambda\gamma$, while the quantum correction to the mass of the excitations in the Z_2 model is:

$$\delta m_\psi^2 = 2\delta m^2 = 12\lambda\gamma \qquad (4.24)$$

In the sine-Gordon model (Eq. (1.51))

$$V_R = \frac{\alpha}{\beta^2}[1 - e^{-\gamma\beta^2}\cos(\beta\phi)] + \epsilon_0 \qquad (4.25)$$

and the quantum corrections to the parameters can be read off.

Returning to the general expression in Eq. (4.21), the bare parameters occurring in $V(\phi)$ can be chosen to absorb the cut-off dependent factors. Then the potential V_R is given entirely in terms of finite physical parameters. If the classical solution is found for the physical value of the coupling constant, denoted by λ_R, then ΔE_{cl} only depends on the correction to the mass term, δm_ψ^2,

$$\Delta E_{\text{cl}} = \frac{E_{\text{cl}}'[m_{\psi,R}; \lambda_R]}{2m_{\psi,R}}\delta m_\psi^2 \qquad (4.26)$$

where E_{cl}' denotes derivative of E_{cl} with respect to the mass, $m_{\psi,R}$. At this stage, we are still left with the last two terms in Eq. (4.20) involving the phase shifts. However, there is no general prescription for finding the phase shifts, and each problem has to be dealt with individually.

Equation (4.20) is our final general expression for the ground state energy of the quantized kink provided that the classical kink solution gives rise to a reflectionless potential. To make further progress one needs to find $E_{\text{cl}}[m_{\psi,R}; \lambda_R]$, ΔE_{cl}, ω_b, and the derivative of $\alpha(k)$. These quantities are model specific and we shall find them in the $\lambda\phi^4$ and sine-Gordon models in the next two sections.

Before going on to some examples, it is helpful to track the coupling constant dependence of the various terms in Eq. (4.16). We write the potential as

$$V(\phi) = -\frac{m^2}{2}\phi^2 + \epsilon S(\phi) \tag{4.27}$$

where m is the mass parameter, ϵ is the small coupling constant, and S is some unspecified function of ϕ, perhaps containing other parameters. The classical energy term in Eq. (4.16) is inversely proportional to the coupling constant. So the leading order correction is independent of the coupling constant. In the second term, δm_ψ^2 is proportional to the coupling constant but E'_{cl} is inversely proportional to the coupling constant. Hence the product is independent of the coupling constant. Next we come to the coupling constant dependence of the energy eigenvalues and the phase shifts. The spectrum of excitations is found by solving for eigenmodes in the kink background. The kink background provides a potential with which the excitations interact. The important point here is that this potential is non-trivial even to zeroth order in the coupling constant. The vacuum expectation value of ϕ, denoted by ϕ_0, is found by minimizing V. Therefore

$$\frac{S'(\phi_0)}{\phi_0} = \frac{m^2}{\epsilon} \tag{4.28}$$

Then

$$V''(\phi_0) = -m^2 + \epsilon S''(\phi_0) \tag{4.29}$$

and approximating $S''(\phi_0)$ as $S'(\phi_0)/\phi_0$,

$$V''(\phi_0) \sim -m^2 + \epsilon \frac{S'(\phi_0)}{\phi_0} \sim m^2 \tag{4.30}$$

Hence the scattering potential in Eq. (4.3) for the mode functions is independent of the coupling constant, and the phase shifts, $\alpha(k)$, are non-trivial even to zeroth order in the coupling constant.

As we see in the specific examples given below, both δm_ψ^2 and the last sum in Eq. (4.16) are divergent. However, the divergences cancel, leading to a finite result for the energy.

4.2 Example: Z_2 kink

We now find the energy of the quantized Z_2 kink by evaluating explicitly the terms in Eq. (4.16).

The classical energy piece is already known from Eq. (1.20)

$$E_{cl}[m_{\psi,R}; \lambda_R] = \frac{2\sqrt{2}}{3}\frac{m^3}{\lambda_R} = \frac{m_{\psi,R}^3}{3\lambda_R} \tag{4.31}$$

Then ΔE_{cl} is given by Eq. (4.26) and E'_{cl} is found by differentiating Eq. (4.31)

$$E'_{cl}[m_{\psi,R}; \lambda_R] = \frac{m^2_{\psi,R}}{\lambda_R} \qquad (4.32)$$

The mass correction δm^2_ψ arises owing to the interaction term $\lambda\phi^4/4$ in this model. The calculation of δm^2_ψ is quite involved since it requires renormalization in a model with spontaneous symmetry breaking, which means that we should find the mass correction from the action in Eq. (1.5). Then there are both cubic and quartic interactions. This calculation can be found in quantum field theory textbooks, for example [119]. The end result is

$$\delta m^2_\psi = \frac{3\lambda_R}{2\pi} \int \frac{dk}{\sqrt{k^2 + m^2_{\psi,R}}} \qquad (4.33)$$

The integral in Eq. (4.33) is divergent. However it is only one term in the expression for the quantum kink energy in Eq. (4.18). In particular, the last term with the phase shifts is also divergent, but the quantum kink energy is finite since the two divergences cancel. Note that we can replace m_ψ by $m_{\psi,R}$ in the final integral since we are only evaluating the leading order correction.

Next consider the terms in Eq. (4.16) that involve the spectrum of fluctuations about the classical kink. To find the spectrum, substitute Eq. (4.1) in the field equation, Eq. (1.4), and expand to lowest non-trivial order in ψ. This was already done in Section 3.2 and we now summarize the results

$$\omega_0 = 0, \qquad \chi_0 = \text{sech}^2 z$$
$$\omega_1 = \frac{\sqrt{3}}{2}m_\psi, \qquad \chi_1 = \sinh z\, \text{sech}^2 z$$
$$m_\psi < \omega < \infty, \qquad \chi_k = e^{ikx}[3\tanh^2 z - 1 - w^2 k^2 - i\,3wk\,\tanh z]$$

where $z = x/w = m_\psi x/2$, and the dispersion relation is

$$\omega_k^2 = k^2 + m^2_\psi \qquad (4.34)$$

Note that the eigenvalues ω_k are independent of the coupling constant because λ does not occur in Eq. (3.10) if $m_\psi = \sqrt{2\lambda}\eta$ is held fixed. However, this statement is only true to leading order in λ because the mass parameter, the kink width, and indeed the form of the kink solution get modified owing to quantum corrections, and induce λ dependence in the spectrum. Since we are only working to leading order in quantum corrections, the mass parameter m_ψ entering Eq. (3.10) and the definition of the kink width, w, are the same as $m_{\psi,R}$.

The next step is to impose periodic boundary conditions with period $L \to \infty$ on the scattering state. For this we find the asymptotic behavior of χ_k

$$\chi_k \to e^{ikz}(2 - w^2k^2 \mp i\, 3wk) \propto \exp\{i(kz \pm \alpha(k)/2)\}, \quad z \to \pm\infty \quad (4.35)$$

from which the phase shifts follow

$$\alpha(k) = 2\tan^{-1}\left[\frac{3wk}{w^2k^2 - 2}\right] \quad (4.36)$$

Hence

$$\frac{d\alpha}{dk} = -6w\frac{w^2k^2 + 2}{(w^2k^2 + 1)(w^2k^2 + 4)} \quad (4.37)$$

and

$$\Delta(k) \to \frac{6}{wk}, \quad |k| \to \infty \quad (4.38)$$

Now we combine all the terms in Eq. (4.20)

$$E = \frac{m_{\psi,R}^3}{3\lambda_R} + \frac{\sqrt{3}}{4}m_{\psi,R} - \frac{3}{2\pi}m_{\psi,R} - \frac{3m_{\psi,R}^3}{16\pi}\int_{-\infty}^{+\infty}\frac{dk}{\sqrt{k^2 + m_{\psi,R}^2}}\frac{1}{k^2 + m_{\psi,R}^2/4} \quad (4.39)$$

The last integral is done easily yielding the final result for the kink mass with leading order quantum correction

$$E = \frac{m_{\psi,R}^3}{3\lambda_R} - \left(\frac{3}{\pi} - \frac{1}{2\sqrt{3}}\right)\frac{m_{\psi,R}}{2}$$

$$= \frac{m_{\psi,R}^3}{3\lambda_R} - 0.33m_{\psi,R} \quad (4.40)$$

Note the minus sign in front of the quantum correction to the energy. In Section 4.5 we show that this is a general feature.

4.3 Example: sine-Gordon kink

To quantize the sine-Gordon kink of Section 1.9, we follow the same procedure as for the Z_2 kink. The mode functions now satisfy

$$-\frac{d^2\psi}{dX^2} + (2\tanh^2 X - 1)\psi = \frac{\omega^2}{m_\psi^2}\psi \quad (4.41)$$

where $X \equiv m_\psi x$. The kink solution, from Eq. (1.52), is

$$\phi_k = \frac{4}{\beta} \tan^{-1} \left(e^{\sqrt{\alpha} x} \right) \equiv \frac{4 m_\psi}{\sqrt{\lambda}} \tan^{-1} \left(e^{m_\psi x} \right) \tag{4.42}$$

where $\lambda \equiv \alpha \beta^2$. The classical energy (Eq. (1.55)) is

$$E_{\text{sG,cl}} = 8 \frac{\sqrt{\alpha}}{\beta^2} \equiv 8 \frac{m_\psi^3}{\lambda} \tag{4.43}$$

The spectrum has only one bound state, the translational zero mode given by

$$\omega_1 = 0, \qquad \psi_0 = \frac{d\phi_k}{dx} = \frac{2m_\psi^2}{\sqrt{\lambda}} \text{sech}(m_\psi x) \tag{4.44}$$

The scattering state with wave-vector k can be written quite generally in terms of hypergeometric functions (see [113], Vol. II, Section 12.3, or Appendix C)

$$\psi_\kappa = N (\cosh X)^{i\kappa X} F \left(-i\kappa - 1, -i\kappa + \frac{1}{2} + \frac{3}{2} |1 - i\kappa| \frac{e^{-X}}{e^X + e^{-X}} \right) \tag{4.45}$$

where N is a normalization factor and $\kappa = k/m_\psi$ corresponds to the wave-vector. The phase shifts are found by taking the asymptotic forms of Eq. (4.45)

$$\psi_\kappa \to N e^{ikX}, \qquad X \to \infty$$
$$\to N e^{i(\pi + 2\theta)} e^{ikX}, \qquad X \to -\infty \tag{4.46}$$

where $\tan \theta = \kappa$. Hence the phase shift is

$$\alpha_k = \pi - 2 \tan^{-1} \left(\frac{k}{m_\psi} \right) \tag{4.47}$$

Therefore

$$\frac{d\alpha_k}{dk} = \frac{-2m_\psi}{k^2 + m_\psi^2} \tag{4.48}$$

At large $|k|$, $\Delta(k)$ (as needed in Eq. (4.20)) is given by

$$\Delta(k) = \frac{2m_\psi}{k}, \qquad |k| \to \infty \tag{4.49}$$

To find ΔE_{cl} occurring in Eq. (4.20), we can use the renormalized potential in Eq. (4.25). The parameter β, which occurs in the argument of the cosine function, is taken to be the physical (renormalized) value, while

$$(\sqrt{\alpha})_b = (\sqrt{\alpha})_R \left(1 + \frac{\beta^2}{2} \gamma \right) \tag{4.50}$$

to leading order in β^2. The subscripts refer to bare and renormalized quantities and γ is defined in Eq. (4.23). Therefore

$$\Delta E_{\text{cl}} = \frac{m_{\psi,\text{R}}}{2\pi} \int_{-\Lambda}^{+\Lambda} \frac{dk}{\sqrt{k^2 + m_\psi^2}} \tag{4.51}$$

Finally, with $\sum \omega_b = 0$, we can put together all the various terms in Eq. (4.20) to get

$$E = \frac{8m_{\psi,\text{R}}^3}{\lambda_\text{R}} + \frac{m_{\psi,\text{R}}}{2\pi} \int_{-\Lambda}^{+\Lambda} \frac{dk}{\sqrt{k^2 + m_\psi^2}} + 0 - \frac{m_{\psi,\text{R}}}{\pi} - \frac{m_{\psi,\text{R}}}{2\pi} \int_{-\Lambda}^{+\Lambda} \frac{dk}{\sqrt{k^2 + m_\psi^2}}$$

$$= \frac{8m_{\psi,\text{R}}^3}{\lambda_\text{R}} - \frac{m_{\psi,\text{R}}}{\pi}$$

$$= \frac{8m_{\psi,\text{R}}^3}{\lambda_\text{R}} - 0.32 m_{\psi,\text{R}} \tag{4.52}$$

Once again the quantum correction is negative and, coincidentally, quite close to the Z_2 value (see Eq. (4.40)).

4.4 Quantized excitations of the kink

So far we have only calculated the quantum correction to the mass of the kink in its ground state. Now consider the excited states of the kink.

As in the second quantization of a free quantum field theory, particle creation and annhilation operators are introduced for each of the excitation modes of the kink. As we shall see, this is straightforward except for the zero mode. The end result is a procedure for doing quantum field theory with both particles and kinks included in the spectrum of states. Here we only give some introductory remarks. For a more extended discussion see [67, 126].

Let us denote the bound state mode functions by $F_b(t, x)$ and the scattering mode functions by $f_k(t, x)$. The t dependence is of the form $\exp(-i\omega_i t)$ where ω_i is the frequency of the bound or scattering mode. Then the second quantized field operator is

$$\phi = \phi_\text{k}(x) + \sum_b \left[\hat{c}_b F_b(t, x) + \hat{c}_b^\dagger F_b^*(t, x) \right] + \sum_k \left[\hat{a}_k f_k(t, x) + \hat{a}_k^\dagger f_k^*(t, x) \right] \tag{4.53}$$

where ϕ_k is the classical kink solution, $c_b^\dagger$, c_b are creation/annihilation operators for the bound states, and similarly $a_k^\dagger$, a_k are creation/annihilation operators for the scattering states. Now, for the zero mode, $\omega = 0$ and $F_0(t, x) = F_0^*(t, x)$. Therefore the zero mode contribution to the sum is

$$[\hat{c}_0 + \hat{c}_0^\dagger] F_0(x) \tag{4.54}$$

Since c_0 and $c_0^\dagger$ are only present in the combination $c_0 + c_0^\dagger$ let us define $b_0 = \hat{c}_0 + \hat{c}_0^\dagger$ which is then the annihilation operator for the zero mode. However, note that $b_0^\dagger = b_0$ and so $[b_0, b_0^\dagger] = [b_0, b_0] = 0$: the zero mode is classical as the operator b_0 commutes with all other operators. This is to be contrasted with $[a_k, a_p^\dagger] = 2\pi \delta(k - p)$.

Just as the translation mode is a bosonic zero mode, there can also be fermionic zero modes that we discuss in Chapter 5. In that case, the creation and annihilation operators satisfy anticommutation relations leading to $\{b_0, b_0^\dagger\} = 0$. This relation has the remarkable consequence of leading to fractional quantum numbers as we discuss in Chapter 5.

4.5 Sign of the leading order correction

A striking feature of the leading order quantum corrections to the energies of the Z_2 and sine-Gordon kink is that they are negative. In other words, quantum effects reduce the mass of the kink. A variational argument [104] (Coleman, S., 1992, private communication) shows that this observation holds true quite generally in one dimension.[1]

Let the Hamiltonian of the $1 + 1$ dimensional system be

$$H \equiv \int dx \mathcal{H} = \int dx \, [\mathcal{H}_0 + V(\phi)] \tag{4.55}$$

where ϕ is a scalar field,

$$\mathcal{H}_0 \equiv \frac{1}{2}\pi^2 + \frac{1}{2}(\partial_x \phi)^2 \tag{4.56}$$

and π is the canonical field momenta. Written in this way, the parameters entering the Hamiltonian are bare parameters and subject to renormalization. In one spatial dimension, however, it can be shown that [35]

$$\mathcal{H} = N_m [\mathcal{H}_0 + V_R] \tag{4.57}$$

where N_m denotes normal ordering with respect to free particles of mass m, and the renormalized potential is (Eq. (4.21))

$$V_R = \exp\left\{ \frac{1}{8\pi} \left(\ln \frac{4\Lambda^2}{m^2} \right) \frac{d^2}{d\phi^2} \right\} V(\phi) + \epsilon \tag{4.58}$$

---

[1] The conclusion may not hold if the model also contains fermionic fields.

where Λ is an ultraviolet momentum cut-off and ϵ is a constant to be chosen such that $\langle 0|H|0 \rangle = 0$ where $|0\rangle$ is the true ground state of the model.

The energy of the kink, including the contribution of quantum fluctuations in the ground state, is

$$E = {}_k\langle 0|H[\phi_k + \psi]|0\rangle_k \tag{4.59}$$

where $|0\rangle_k$ denotes the vacuum for the quantum fluctuations, ψ, around the classical one kink state ϕ_k.

Straightforward manipulation now gives the quantum correction to the kink mass

$$\begin{aligned}
E - E_{\text{cl,R}} &= {}_k\langle 0|H[\phi_k + \psi] - H[\phi_k]|0\rangle_k \\
&= \int dx \, {}_k\langle 0|N_m(\mathcal{H}_0[\phi_k + \psi] - \mathcal{H}_0[\phi_k] + V_R[\phi_k + \psi] - V_R[\phi_k])|0\rangle_k
\end{aligned}$$

where $E_{\text{cl,R}}$ is the energy of the classical solution obtained with the renormalized potential, V_R. Next we use the variational principle, which states that the ground state energy of a system is minimized in its true ground state, and the expectation of the Hamiltonian in any other trial state gives an upper bound to the ground state energy. If we denote the perturbative vacuum state – the state with zero particles of mass m – by $|0, m\rangle$, then

$$\begin{aligned}
E &\leq E_{\text{cl,R}} + \int dx \, \langle 0, m|N_m(\mathcal{H}_0[\phi_k + \psi] - \mathcal{H}_0[\phi_k] + V_R[\phi_k + \psi] - V_R[\phi_k])|0, m\rangle \\
&= E_{\text{cl,R}}
\end{aligned}$$

The last line follows since there are no ψ independent terms in the expectation value under the integral,[2] and the annihilation operators of ψ occur to the right owing to normal ordering and annihilate the trial vacuum state.

Note that $E_{\text{cl,R}}$ is the energy of the classical solution found by minimizing $H_R[\phi]$, i.e. the Hamiltonian in Eq. (4.55) but with the potential given in Eq. (4.58). Since the true ground state of the system is not known, the constant ϵ is not known either. The potential V_R can be minimized, but there is no guarantee that the minimal value of V_R will be zero. Therefore $E_{\text{cl,R}}$ might get an infinite contribution from integrating $\min(V_R)$ over all of space. Then the variational bound $E \leq E_{\text{cl,R}}$ is not very useful. However, we do know the value of ϵ to lowest order in the coupling constant and this is precisely so that $\langle 0, m|H|0, m\rangle = 0$. This coincides with choosing ϵ so as to make $\min(V_R) = 0$. Hence the bound

$$E \leq E_{\text{cl,R}} = E_{\text{cl}} \tag{4.60}$$

[2] To see this, note that the expectation value vanishes if $\psi = 0$.

where E_{cl} denotes the classical energy without any quantum corrections, is meaningful to leading order in perturbation theory and it provides us with the completely general result that the lowest order correction to the soliton energy is negative.

4.6 Boson-fermion connection

A bosonic field, ϕ, in quantum field theory satisfies the equal time commutation relation

$$[\phi(x, t), \dot{\phi}(y, t)] = \delta(x - y) \tag{4.61}$$

Alternatively, a fermionic field, ψ, satisfies the anticommutation relations

$$\{\psi_a(x, t), \psi_b^{\dagger}(y, t)\} = \delta(x - y)\delta_{ab} \tag{4.62}$$

where $a, b = 1, 2$ label the two components of the spinor in one spatial dimension. It is remarkable that one can construct explicitly a fermionic field ψ satisfying Eq. (4.62) in terms of a bosonic field ϕ that satisfies Eq. (4.61) [108].

The connection between ψ_a and ϕ is

$$\psi_1(x) = C : e^{P_+(x)} :, \quad \psi_2(x) = -iC : e^{P_-(x)} : \tag{4.63}$$

where the c-number C is defined in terms of a mass parameter μ and another cut-off parameter, ϵ,

$$C = \left(\frac{\mu}{2\pi}\right)^{1/2} e^{\mu/8\epsilon} \tag{4.64}$$

The operators $P_{\pm}$ contain a free parameter β and are defined by

$$P_{\pm}(x) = -i\frac{2\pi}{\beta} \int_{-\infty}^{x} d\xi \, \dot{\phi}(\xi) \mp \frac{i\beta}{2}\phi(x) \tag{4.65}$$

The symbol :: in Eq. (4.63) denotes normal ordering with respect to the mass μ. This means that the field ϕ is to be treated as a free field with mass parameter μ and the quantum operator, ϕ, is expanded in terms of creation and annihilation operators that create and destroy particles of this free field theory. A normal ordered operator contains various products of creation and annihilation operators with the annihilation operators always occurring on the right. It is understood that the integral in Eq. (4.65) is cut off at large ξ by a factor $\exp(-\epsilon\xi)$. Note that normal ordering is a symbol and should be treated carefully – normal ordering of strings of operators should be done prior to commuting operators that occur within the string.

To check if Eq. (4.62) is satisfied for $x \neq y$, we use the identity (see Appendix D)[3]

$$: e^{A+B} := e^{-[A^+, B^-]} : e^A :: e^B := e^{-[B^+, A^-]} : e^B :: e^A : \qquad (4.66)$$

where A and B are any two operators that can be written as a linear sum of terms involving only creation or annihilation operators

$$A = A^+ + A^-, \qquad B = B^+ + B^- \qquad (4.67)$$

The commutators $[A^+, B^-]$ and $[B^+, A^-]$ are assumed to be c-numbers. Insertion of Eq. (4.66) in Eq. (4.62) gives the commutation relation in Eq. (4.61) for $x \neq y$.

It is harder to check that the commutation relations in Eq. (4.61) hold when $x = y$. Since products of quantum operators at the same point are singular, the commutator must be evaluated at two different points in space, x and y, followed by the coincidence limit $y \to x$. We now outline the scheme employed in [108].

We want to check the anticommutation relation

$$\{\psi_a(x), \psi_b^\dagger(y)\} = Z\delta(x - y) \qquad (4.68)$$

where the constant Z, possibly infinite, has been introduced in recognition of the fact that the fields get renormalized. Rather than check Eq. (4.68), we can check the equivalent *commutation* relation

$$[j^\mu(x), \psi(y)] = -\left(g^{0\mu} + \frac{\beta^2}{4\pi} \epsilon^{\mu 0} \gamma^5 \right) \psi(x)\delta(x - y) \qquad (4.69)$$

where the current j^μ has been regularized using point-splitting and is defined by

$$j^\mu(x) = \lim_{y \to x} \left\{ \left[\delta^{\mu 0} + \frac{\beta^2}{4\pi}\delta_1^\mu \right][\mu(x - y)]^\sigma \, \bar{\psi}(x)\gamma^\mu \psi(y) + F^\mu(x - y) \right\} \qquad (4.70)$$

where σ is a regularizing parameter and $F^\mu(x - y)$ an unspecified c-valued function. The γ-matrices are defined by the algebra

$$\{\gamma^\mu, \gamma^\nu\} = 2g^{\mu\nu}, \qquad \gamma^5 = i\gamma^0\gamma^1 \qquad (4.71)$$

where $g_{\mu\nu} = \text{diag}(1, -1)$ is the two-dimensional Minkowski metric. An explicit representation of the γ-matrices is given in Eq. (5.15). In Eq. (4.69), $\epsilon^{\mu\nu}$ is the totally antisymmetric tensor.

First the current j^μ is evaluated with ψ_a as given in Eq. (4.63). The evaluation requires

$$[\phi^+(x, t_2), \phi^-(y, t_1)] = \Delta_+[(x - y)^2 - (dt + i\epsilon)^2] \qquad (4.72)$$

---

[3] In the literature it is sometimes incorrectly stated that the identity $e^{A+B} = e^{[B,A]/2}e^A e^B$ (no normal ordering) is being used.

where $dt = t_2 - t_1$ and Δ_+ is the propagator. For small $x - y$

$$\Delta_+ = -\frac{1}{4\pi} \ln[\mu^2\{(x-y)^2 - (dt + i\epsilon)^2\}] + O((x-y)^2) \tag{4.73}$$

By differentiating Eq. (4.72) we can also obtain the commutators of time derivatives of ϕ^+ and ϕ^-. These appear in the evaluation of j^μ since ψ is defined in terms of ϕ in Eq. (4.63).

The result for j^μ is singular in the limit $y \to x$ except for a single choice of the regularizing parameter, σ, occurring in the definition of j^μ. This single choice is

$$\sigma = \frac{\beta^2}{8\pi}\left(1 - \frac{4\pi}{\beta^2}\right)^2 \tag{4.74}$$

With this value of σ, the commutator in Eq. (4.69) can be verified. Thus the ψ operator indeed satisfies the anticommutation relations of a fermionic field. Furthermore, the current can be explicitly calculated, leading to

$$j^\mu = -\frac{\beta}{2\pi}\epsilon^{\mu\nu}\partial_\nu\phi \tag{4.75}$$

To summarize, given a quantum scalar field in $1+1$ dimensions, it is possible to construct a fermionic field from it via the relation (4.63). Starting with a fermionic field, a bosonic field may be constructed from it via Eq. (4.75). Note that the transformations from bosons to fermions and vice versa hold at the quantum operator level and not just at the level of expectation values. Further, they hold for any choice of interactions in the bosonic or the fermionic model. However, in the case when the bosonic model is the sine-Gordon model, the fermionic model obtained by transforming to the fermionic variables is another well-known model, namely the massive Thirring model as we now describe.

4.7 Equivalence of sine-Gordon and massive Thirring models

The sine-Gordon model is given by the Lagrangian (Eq. (1.51))

$$L_{sG} = \frac{1}{2}(\partial_\mu\phi)^2 - \frac{\alpha}{\beta^2}(1 - \cos(\beta\phi)) \tag{4.76}$$

while the massive Thirring model is

$$L_{mT} = i\bar{\psi}\,\partial\!\!\!/\psi - m\bar{\psi}\psi - \frac{g}{2}\bar{\psi}\gamma^\mu\psi\,\bar{\psi}\gamma_\mu\psi \tag{4.77}$$

where ψ is a two-component fermionic field.

In [34] (also see [35]) it is shown that the sine-Gordon model does not have a well-defined ground state for $\beta^2 > 8\pi$. To clarify what this means, consider the simple example of a free field theory

$$L_{\text{free}} = \frac{1}{2}(\partial_\mu \phi)^2 - \frac{\delta}{2}\phi^2 \tag{4.78}$$

This model has a well-defined ground state only in the range $\delta \geq 0$. The model does not have a well-defined ground state for $\delta < 0$. Similarly the sine-Gordon model only has a ground state for a definite range of parameters, though the reasons are much more subtle.[4] The sine-Gordon only has a well-defined ground state if the parameter β^2 is restricted to lie in the interval $(0, 8\pi)$.

In the range, $0 \leq \beta^2 \leq 8\pi$, there is a one-to-one mapping between vacuum expectation values of a string of operators in the sine-Gordon model to those in the massive Thirring model. This means that any vacuum expectation value in the sine-Gordon model has a "corresponding" vacuum expectation value in the massive Thirring model. This strongly suggests that the two models are equivalent, even at the operator level [35].

As we have seen in the last section, there is indeed a two-component fermionic field, ψ, that can be constructed from a bosonic field ϕ (Eq. (4.63)). In [108] it was shown that ψ also obeys the equations of motion of the massive Thirring model if the bosonic field ϕ obeys the equations for the sine-Gordon equation with the coupling constant g written in terms of the coupling constant β as

$$\frac{g}{\pi} = 1 - \frac{4\pi}{\beta^2} \tag{4.79}$$

Note that when the sine-Gordon model is weakly coupled (small β), the massive Thirring model is strongly coupled and vice versa. Hence the sine-Gordon model and the massive Thirring model are completely equivalent as quantum field theories but one is a better description at small β (large g) and the other at large β (small g).

What has the equivalence of the sine-Gordon and massive Thirring models got to do with kinks? Consider the commutation relations between ϕ and ψ. Using Eq. (4.63) and the identity (see Appendix D) $[A, : e^B :] =: [A, e^B] :$ with $A = \phi(y)$ and $: e^B := \psi$ we find

$$[\phi(y), \psi(x)] = \frac{2\pi}{\beta}\psi(x), \quad (y < x) \tag{4.80}$$

$$[\phi(y), \psi(x)] = 0, \quad (y > x) \tag{4.81}$$

Now consider the action of $\psi(x)$ on an eigenstate, $|s\rangle$ of the field operator ϕ. Let us choose this eigenstate to be such that

$$\phi|s\rangle = 0 \tag{4.82}$$

---

[4] For example, in contrast to the model in Eq. (4.78), the *classical* sine-Gordon model has well-defined global minima for all values of the coupling constant β.

If we write $|s'\rangle = \psi(x)|s\rangle$, the relation in Eq. (4.80) gives

$$\phi(y)|s'\rangle = \frac{2\pi}{\beta}|s'\rangle, \quad (y < x) \tag{4.83}$$

and Eq. (4.81) gives

$$\phi(y)|s'\rangle = 0, \quad (y > x) \tag{4.84}$$

Therefore the state obtained after action by $\psi(x)$ is one where the value of ϕ is $2\pi/\beta$ for $y < x$ and 0 for $y > x$. In other words, the field $\psi(x)$ creates a step-change of ϕ. The step-function profile is viewed as a "bare kink" which gets dressed by quantum effects that produce a smooth kink profile with some finite width. So the field $\psi(x)$ is the creation operator for a (bare) soliton at location x. In the Thirring model, the field $\psi(x)$ is interpreted as the creation operator for a fermion located at x. Hence the sine-Gordon kink is identified with the fermion in the massive Thirring model.

The topological charge on a sine-Gordon kink is

$$Q_k = \int dx\, j_B^0 = \int dx\, j_F^0 \tag{4.85}$$

where the fermionic current is defined in terms of the bosonic current in Eq. (4.75). Therefore the fermion in the massive Thirring model carries the topological charge of the sine-Gordon kink. In other words, the kink of the strongly coupled sine-Gordon model is better described as a weakly coupled fermion of the massive Thirring model. Here we see the duality between particles and solitons.

Can we also interpret the bosonic particles of the sine-Gordon model in terms of "solitons" of the massive Thirring model? The massive Thirring model only contains fermions, and classical solutions of the Dirac equation do not have the interpretation of solitons. This is because the fermionic fields anticommute and fermions obey the Pauli exclusion principle. Instead a classical solution of the Dirac equation is a state that one (and only one) fermion can occupy. However, there can be bound states of two or more fermions since the force between a fermion and an antifermion is attractive for $g > 0$. A bound state of two fermions can be shown to correspond to a particle of the sine-Gordon field ϕ. If the fermions in the weakly coupled massive Thirring model have mass m, then the bound state energy is approximately $2m$ since it involves two fermions. However, the binding energy decreases (becomes more negative) with increasing interaction strength, g, and eventually the bound state becomes lighter than a single fermion. At this stage, a suitable description of the system is in terms of the bound state being the fundamental degree of freedom as in the sine-Gordon model.

The bound state of two massive Thirring fermions is also a bound state of two sine-Gordon kinks i.e. a breather. Hence it should be possible to interpret the breather as a particle in the sine-Gordon model. This is seen to be true when the breather is quantized [38–41, 35]. Then, to lowest order, the energy levels of the quantized breather are equal to the mass of one, two, three, etc. particles of the sine-Gordon particle.

4.8 Z_2 kinks on the lattice

Lattice field theory provides another tool to probe the quantum nature of solitons and, in particular, the variation of mass with coupling constant.

The starting point is the action for the Z_2 model defined in Eq. (1.2). The action is to be inserted in the Feynman path integral, which can then be used to find expectation values for any quantum operator. In the Feynman path integral, it is necessary to integrate over field configurations, and this is done numerically on a discretized Euclidean space-time. The reader is referred to the lattice literature for details [37, 112, 141]. Here we shall give the results relevant to the Z_2 kink.

The mass of a Z_2 kink is defined as the expectation value of a suitable operator defined on the lattice in the limit that the lattice spacing, a, goes to zero. One important issue is that there are several different candidate operators on the lattice that all go to the correct limit as $a \rightarrow 0$ and, in practice, it is not possible to take the limit all the way to $a = 0$. At best, the numerical analysis gives the expectation of the operator on the lattice for several different values of a and then some scheme must be found for extrapolating the results to $a \rightarrow 0$. In [32], the authors evaluate the mass of the Z_2 kink using two different lattice operators. The results are shown in Fig. 4.2. We note that the kink mass decreases monotonically as the coupling constant increases and remains bounded by the classical mass. At a certain coupling, the kink mass goes to zero, and the kink, not the ϕ quanta, is the lightest degree of freedom in the model.

The mass of the sine-Gordon kink has been calculated analytically for a range of parameters in [156] (also see [35]).

4.9 Comments

Several researchers have taken alternate paths to studying quantized kinks. In supersymmetric theories there is greater control over quantum corrections and the mass can, in some cases, be evaluated exactly [51]. Alternate methods to quantize supersymmetric kinks have also been developed in [66]. Variational methods to study the $\lambda \phi^4$ theory have been developed in [49]. The scattering of kinks in classical and quantum theory has been studied in [153]. Kink masses and scattering have also been calculated in [132] using the Hartree approximation.

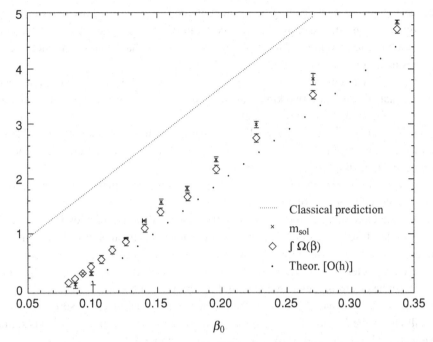

Figure 4.2   The figure shows how the mass of the Z_2 kink depends on the inverse coupling constant, $\beta_0 = 1/\lambda_0$, where $\lambda_0 \equiv 6\lambda a^2$ is the coupling constant in the discrete theory on a 48×48 lattice and a is the lattice spacing. (The factor of 6 is due to our choice of $1/4$ in the λ term in Eq. (1.2) as opposed to $1/4!$ in [32].) The lattice mass parameter, $r_0 \equiv -m^2 a^2$, is held fixed at $r_0 = -2.2$. From the plot we see that the classical value of the kink mass is larger than the quantum value. The one-loop corrected mass (see Section 4.2) and the mass found by using two different choices of the lattice mass operator are also shown. The kink mass vanishes at $\beta_0 = 0.0804$. [Figure reprinted from [32].]

The construction of fermion operators from boson operators and vice versa has been discussed and used extensively in condensed matter applications under the name of bosonization. A review, in addition to an historical introduction, may be found in [44]. Finally, the technique of bosonization has also been applied to thermal systems in [69].

4.10 Open questions

1. The quantum corrections to the Z_2 and sine-Gordon kinks were calculated explicitly using the phase shifts. However, the phase shift approach only works if the potential $U(x)$ is reflectionless. What are the conditions necessary for a potential to be reflectionless? Are reflectionless potentials always in factorizable form (see Section 3.3)? (The example of a step-function potential shows that the converse is not true.)

2. We have shown that the leading order quantum correction to the kink mass is always negative. Can this statement be generalized to all orders? Can one show that the mass of a kink goes to zero in the strong coupling limit? Or perhaps that it is monotonically decreasing as a function of increasing coupling constant?

3. If the Z_2 kink at strong coupling is to be viewed as a particle, then the particle must obey unusual statistics because two kinks cannot be next to each other. Discuss this statistics and its implications for the dual model.

4. From the $SU(5) \times Z_2$ example we learned that a classical kink may be embedded in many different ways in "large" models. On quantization, do the different embeddings correspond to distinct degrees of freedom?

5. Does the addition of fermionic particles change the conclusion that quantum corrections always reduce the energy of a kink?

6. For the sine-Gordon model we have explicitly seen that there is a relation between kinks and particles. It seems reasonable that the connection holds in other models too. In $3 + 1$ dimensions, we could expect the connection to exist between magnetic monopoles and observed particles (e.g. [162, 103]). Construct a model that has families of solitons, similar to the electron, muon, and tau families observed in Nature (see [122]).

7. In Section 3.1 we have discussed the existence of quasi-breather solutions called "oscillons" in the Z_2 model. Can quantum oscillons have an interpretation that is similar to quantized breathers as discussed at the end of Section 4.7?

5

Condensates and zero modes on kinks

In this chapter we study the effect of a kink on other bosonic or fermionic fields that may be present in the system. Under certain circumstances, it might be energetically favorable for a bosonic field, denoted by χ, to become non-trivial within the kink. Then we say that there is a "bosonic condensate" which is trapped on the kink. On a domain wall, the condensate has dynamics that are restricted to lie on the world-sheet of the wall.

The situation is similar for a fermionic field though there are subtleties. For a fermionic field, denoted by ψ, the Dirac equation is solved in the presence of a kink background made up of bosonic fields. This determines the various quantum modes that the fermionic excitations can occupy. In several cases, there can be "zero modes" of fermions in the background of a kink and this leads to several new considerations. (Fermionic zero modes were first discovered in [27, 84] in the context of strings.) In addition to the zero mode, there may be fermionic bound states. The high energy states that are not bound to the wall are called "scattering (or continuum) states."

A difference between bosonic and fermionic condensates is that bosonic solutions can be treated classically but fermionic solutions can only be interpreted in quantum theory. For example, while there may be a bosonic solution with $\chi = 0$, the solution $\psi = 0$ of the Dirac equation has no meaning because this solution is not normalizable. Solutions of the Dirac equation are only meant to supply us with the modes that fermionic particles or antiparticles can occupy, and as such are required to be normalizable. It is a separate issue to decide if the modes are occupied or not. A mode contributes to the energy of the soliton only if it is occupied. This is quite different from the bosonic case in which there can be a classical condensate, on top of which there are modes that may or may not be occupied by bosonic particles. Fermions can form a classical condensate only after they have paired up to form bosons ("Cooper pairs"), and this leads to superfluidity or superconductivity.

Fermionic zero modes can lend solitons some novel properties such as fractional quantum numbers (see Section 5.3).

5.1 Bosonic condensates

Consider the model

$$L = L_k[\phi] + \frac{1}{2}(\partial_\mu \chi)^2 - U(\phi, \chi) \tag{5.1}$$

where $L_k[\phi]$ is the Lagrangian that leads to a kink solution in ϕ. For example, L_k can be the Lagrangian for the Z_2 or sine-Gordon models discussed in Chapter 1. χ is another scalar field that interacts with ϕ via some general interaction potential $U(\phi, \chi)$. Note that $U(\phi, \chi)$ does not contain any terms that are independent of χ – those are included in the potential, $V(\phi)$, that occurs in L_k. As an example, we could have

$$U(\phi, \chi) = -\frac{m_\chi^2}{2}\chi^2 + \frac{\lambda_\chi}{4}\chi^4 + \frac{\sigma}{2}\phi^2\chi^2 \tag{5.2}$$

We are assuming that the parameters in the model are chosen so that the minimum of the full potential, $V + U$, is at $\phi \neq 0$ but $\chi = 0$. This requirement also excludes terms that are linear in χ (e.g. $\phi^2\chi$).

In the fixed background of the kink, χ satisfies the classical equation of motion

$$\partial_t^2 \chi - \partial_x^2 \chi + U_\chi(\phi_k(x), \chi) = 0 \tag{5.3}$$

where U_χ denotes the derivative of U with respect to χ and ϕ_k is the kink solution. Far from the wall, the lowest energy solution is $\chi(\pm\infty) = 0$.

A solution to Eq. (5.3) is $\chi(x) = 0$ and the energy of this solution is equal to the kink energy in the model L_k. However, the trivial solution may not be the one of lowest energy. To show that a lower energy solution exists, we need only show that the trivial solution, $\chi = 0$, is unstable. Then we consider linearized perturbations of the form $\chi = \cos(\omega t) f(x)$ around the trivial solution. Inserting this form into Eq. (5.3) leads to the Schrödinger equation

$$-\partial_x^2 f + U_{\chi\chi}(\phi_k(x)) f = \omega^2 f \tag{5.4}$$

where $U_{\chi\chi}$ denotes the second derivative of V with respect to χ. If this equation has solutions with $\omega^2 < 0$, it implies that there are solutions for χ on the kink background that grow with time as $\cosh(+|\omega|t)$, denoting an instability of the state with $\chi = 0$. This means that the solution with least energy must have a non-trivial χ configuration. The lowest energy χ configuration is non-zero inside the kink and vanishing outside and is called a "bosonic condensate" (or simply "condensate").

5.1.1 Bosonic condensate: an example

A simple example in which there is a bosonic condensate on a Z_2 kink can be found in the model of Eq. (5.1), or explicitly,

$$L = \frac{1}{2}(\partial_\mu \phi)^2 + \frac{1}{2}(\partial_\mu \chi)^2 - \frac{\lambda}{4}(\phi^2 - \eta^2)^2 + \frac{m_\chi^2}{2}\chi^2 - \frac{\lambda_\chi}{4}\chi^4 - \frac{\sigma}{2}\phi^2\chi^2 \qquad (5.5)$$

Ignoring the condensate field χ, the kink solution is

$$\phi_k = \eta \tanh\left(\frac{x}{w}\right) \qquad (5.6)$$

and the Schrödinger equation corresponding to Eq. (5.4) is

$$-\partial_x^2 f + \left[-m_\chi^2 + \sigma\eta^2 \tanh^2 X\right]f = \omega^2 f \qquad (5.7)$$

where $X = x/w$.

With $\sigma\eta^2 > m_\chi^2$, we see that the Schrödinger potential is asymptotically positive, and hence $f(\pm\infty) = 0$. This is consistent with the requirement that χ not have a vacuum expectation value. At the origin, $U_{\chi\chi} < 0$, and hence the Schrödinger potential is a well that is centered at the origin. Since a potential well in one dimension always has a bound state [139], it follows that there is at least one bound state for χ. For a deep enough well i.e. large enough m_χ^2, the bound state has negative energy eigenvalue ($\omega^2 < 0$), and the trivial solution $\chi = 0$ is unstable. Hence there is a range of parameters (m_χ^2) for which a χ condensate exists.

To determine the range of m_χ^2 for which there is an instability, consider the critical case when there is a zero eigenvalue solution, f_0, of Eq. (5.7). Then we can write

$$-\partial_x^2 f_0 + \frac{\sigma\eta^2}{3}[3\tanh^2 X - 1]f_0 = \left[m_\chi^2 - \frac{\sigma\eta^2}{3}\right]f_0 \qquad (5.8)$$

This is exactly the same form as Eq. (3.8), together with the potential in Eq. (3.10), provided we identify 3λ with σ, and ω^2 with the term within square brackets on the right-hand side. Since the lowest energy eigenvalue is zero for Eq. (3.8), there is a zero eigenvalue for Eq. (5.7) if

$$m_\chi^2 = \frac{\sigma\eta^2}{3} \qquad (5.9)$$

For a larger value of m_χ^2, Eq. (5.7) has a negative eigenvalue, signaling an instability. Therefore a condensate solution exists in the range

$$\frac{\sigma\eta^2}{3} < m_\chi^2 < \sigma\eta^2 \qquad (5.10)$$

The exact profile for the condensate can be found by solving the full coupled equations of motion for ϕ and χ. This includes the non-linear terms in χ and the back-reaction of the condensate on the kink, and, in most cases, has to be done numerically. Let us denote the solution obtained in this way by $(\phi_k(x), \chi_0(x))$. Then

$$\chi(t, x, y, z) = \chi_0(x)\cos(\omega t - k_y y - k_z z + \theta_0), \quad \omega = \sqrt{k_y^2 + k_z^2} \tag{5.11}$$

where θ_0 is a constant, is also a solution. The reason is simply that

$$\left(\partial_t^2 - \partial_y^2 - \partial_z^2\right)\cos(\omega t - k_y y - k_z z + \theta_0) = 0 \tag{5.12}$$

The trigonometric form of the solution in Eq. (5.11) was chosen so as to obtain a real solution. An identical analysis in the case when χ is a complex field leads to

$$\chi(t, x, y, z) = \chi_0(x)e^{\pm i(\omega t - k_y y - k_z z + \theta_0)} \tag{5.13}$$

The solution represents waves propagating in the (k_y, k_z) direction in the plane of the domain wall.

5.2 Fermionic zero modes

Fermionic fields may be coupled to the kink via terms that respect the discrete symmetries in the bosonic sector that are responsible for the existence of the kink. In the case of the Z_2 model, the coupling can be a Yukawa term and the Lagrangian may be written as

$$L = L_\phi + i\bar\psi\,\partial\!\!\!/\psi - g\phi\bar\psi\psi \tag{5.14}$$

where L_ϕ denotes the scalar part of the Lagrangian, $\partial\!\!\!/ \equiv \gamma^\mu \partial_\mu$, g is the coupling constant, ψ is a four-component fermionic field, and γ^μ are the Dirac matrices that satisfy $\{\gamma^\mu, \gamma^\nu\} = 2g^{\mu\nu}$ with $g_{\mu\nu} = \mathrm{diag}(1, -1, -1, -1)$ being the space-time metric. The explicit representation of the Dirac matrices that we adopt is

$$\gamma^0 = \begin{pmatrix} 0 & 1 \\ 1 & 0 \end{pmatrix}, \quad \gamma^i = \begin{pmatrix} 0 & \sigma^i \\ -\sigma^i & 0 \end{pmatrix} \tag{5.15}$$

where $i = 1, 2, 3$ (also sometimes written as $i = x, y, z$) and the Pauli spin matrices are defined as

$$\sigma^1 = \begin{pmatrix} 0 & 1 \\ 1 & 0 \end{pmatrix}, \quad \sigma^2 = \begin{pmatrix} 0 & -i \\ i & 0 \end{pmatrix}, \quad \sigma^3 = \begin{pmatrix} 1 & 0 \\ 0 & -1 \end{pmatrix} \tag{5.16}$$

The Yukawa interaction term in the model in Eq. (5.14) respects the $\phi \to -\phi$ symmetry of the Z_2 model provided we also transform the fermion field by

$\psi \to \psi' = \gamma^5 \psi$ where

$$\gamma^5 \equiv i\gamma^0\gamma^1\gamma^2\gamma^3 = \begin{pmatrix} -1 & 0 \\ 0 & 1 \end{pmatrix} \tag{5.17}$$

This can be seen by using the properties $(\gamma^5)^\dagger = \gamma^5$, $\{\gamma^5, \gamma^\mu\} = 0$ and $(\gamma^5)^2 = 1$.[1]

If $\phi_k(x)$ denotes the kink solution, the Dirac equation in the kink background is

$$i\partial\!\!\!/\psi - g\phi_k(x)\psi = 0 \tag{5.18}$$

Let us first try and solve Eq. (5.18) explicitly. Recognizing that ϕ_k does not depend on t, y, and z, we write the ansatz

$$\psi = f(t, y, z)\xi(x) \tag{5.19}$$

where $f(t, y, z)$ is a function while $\xi(x)$ is a four-component spinor. With this ansatz, the Dirac equation separates

$$i\gamma^a \partial_a f = -\gamma^a k_a f \tag{5.20}$$

$$i\gamma^x \partial_x \xi - g\phi_k \xi = +\gamma^a k_a \xi \tag{5.21}$$

where $\gamma^a k_a$ is the constant matrix of separation and the index a runs over t, y, z. Requiring that the fermion be localized on the wall, we get the boundary conditions

$$\xi(\pm\infty) = 0 \tag{5.22}$$

These boundary conditions are valid only for bound states. If we wish to consider the scattering of fermions off a domain wall, we would choose incoming and reflected plane waves at $x = -\infty$.

The Dirac equations have an infinite number of solutions, corresponding to all the fermion eigenmodes in the domain wall background. These include fermionic bound states and scattering states. There is one state, however, which is novel because it leads to some very interesting properties of the soliton, described in the sections below. This state is the one with zero energy eigenvalue, also called the "zero mode."

Equation (5.20) can be solved

$$f = \exp(ik_a x^a) \equiv \exp(i(\omega t - k_y y - k_z z)) \tag{5.23}$$

Zero energy is obtained by setting $\omega = 0 = k_y = k_z$ and then $f = 1$. Let us first look at this case ($k_a = 0$).

Multiplying Eq. (5.21) by $i\gamma^x$ we see that $i\gamma^x\xi$ satisfies the same equation of motion as ξ. Therefore if ξ is a solution, then so is $i\gamma^x\xi$. Hence solutions to the

---

[1] The Yukawa term does not respect the $\phi \to \phi + 2\pi/\beta$ symmetry of the sine-Gordon model and hence our discussion of fermion zero modes cannot be used for that case. Nor do we consider the case of fermions with Majorana mass terms [147].

Dirac equation come in distinct pairs unless ξ is an eigenstate of $i\gamma^x$, in which case the two solutions ξ and $i\gamma^x\xi$ are not distinct. The zero mode solution is found by choosing ξ to be an eigenstate of $i\gamma^x$

$$i\gamma^x\xi = c\xi \tag{5.24}$$

and, since $(i\gamma^x)^2 = 1$, we must have $c = \pm 1$. The ξ equation now becomes

$$\partial_x\xi = cg\phi_k\xi \tag{5.25}$$

and the solution is

$$\xi(x) = \xi(0)\exp\left[cg\int_0^x \phi_k(x')dx'\right] \tag{5.26}$$

Assuming $\phi_k(+\infty) > 0$ and $g > 0$, the boundary conditions in Eq. (5.22) are only satisfied if $c = -1$. (The boundary condition at $x = -\infty$ is also satisfied provided $\phi_k(-\infty) < 0$.) Therefore the zero mode solution is

$$\xi(x) = \xi(0)\exp\left[-g\int_0^x \phi_k(x')dx'\right], \quad (g > 0) \tag{5.27}$$

If $\phi_k(+\infty) < 0$ and $g > 0$, the solution is obtained by choosing $c = +1$.

To determine $\xi(0)$, we solve the eigenvalue equation $i\gamma^x\xi(0) = -\xi(0)$ and find

$$\xi(0) = \begin{pmatrix} \alpha \\ \beta \\ i\beta \\ i\alpha \end{pmatrix} \tag{5.28}$$

where α, β are any complex constants. Therefore there are two basis zero modes (with coefficients α and β) and the general zero mode is a linear superposition of these two modes. The constants, α and β, can be fixed by normalizing the wavefunction.

Next consider the case with $k_a \neq 0$. Then Eq. (5.27) is still a solution to Eq. (5.21) provided $k_a\gamma^a\xi(0) = 0$. By explicitly substituting the γ^a matrices and $\xi(0)$, this leads to the two equations

$$k_y\alpha + i(\omega + k_z)\beta = 0 \tag{5.29}$$
$$i(\omega - k_z)\alpha - k_y\beta = 0 \tag{5.30}$$

A solution for α and β exists only if

$$\omega = \pm\sqrt{k_y^2 + k_z^2} \tag{5.31}$$

which is the dispersion relation for a massless particle (see Fig. 5.1). With this relation, the solutions fix the ratio of α and β to obtain

$$\psi = \frac{Ne^{i(\omega t - k_y y - k_z z)}}{2\sqrt{\omega}} e^{-g \int_0^x \phi_k(x')dx'} \begin{pmatrix} \sqrt{\omega + k_z} \\ \mathrm{sgn}(k_y)i\sqrt{\omega - k_z} \\ -\mathrm{sgn}(k_y)\sqrt{\omega - k_z} \\ i\sqrt{\omega + k_z} \end{pmatrix} \qquad (5.32)$$

where N is a normalization constant where $\mathrm{sgn}(k_y) \equiv k_y/|k_y|$.

So far we have not specified the exact form of the kink profile ϕ_k and Eq. (5.32) holds for any model in which the Yukawa interaction term respects the symmetries. Next, as an example, we use the solution for the Z_2 kink (see Eq. (1.9)). Then the integral over ϕ_k can be done explicitly to yield

$$\psi = \frac{Ne^{i(\omega t - k_y y - k_z z)}}{2\sqrt{\omega}} \left[\mathrm{sech}\left(\frac{x}{w}\right)\right]^{g\sqrt{2/\lambda}} \begin{pmatrix} \sqrt{\omega + k_z} \\ \mathrm{sgn}(k_y)i\sqrt{\omega - k_z} \\ -\mathrm{sgn}(k_y)\sqrt{\omega - k_z} \\ i\sqrt{\omega + k_z} \end{pmatrix}, \quad (g > 0) \qquad (5.33)$$

where w is the width of the kink as defined in Eq. (1.21). This is our final expression for the zero mode on the Z_2 kink.

In the asymptotic vacuum, where ϕ is constant, the Dirac equation derived from Eq. (5.14) yields four solutions all with the same momentum. These four states are referred to as spin up and down states for the particle and hole (or antiparticle). On the domain wall, however, there are only *two* zero mode solutions for fixed value of the momentum (k_y, k_z). One of these has positive energy (ω) and the other has negative energy. Therefore the two states may be called particle and hole states but the spin degree of freedom is not present. Consider the special case when $k_y = 0$ and $k_z \neq 0$. Then we have $\omega = \pm k_z$ and the spinor is proportional to $(1, 0, 0, i)^T$ if $\omega = +k_z$, and to $(0, i, -1, 0)^T$ if $\omega = -k_z$. If we also take $k_z = 0$, both these two states have $\omega = 0$ and become degenerate in energy.

The two-fold degeneracy of the zero mode ($\omega = 0$) occurs since we are working in three spatial dimensions where the Dirac spinors have four components. If we find the zero modes in one spatial dimension, the fermions are described by two-component spinors, and then there is only a single zero mode. If we use four-component spinors in one spatial dimension, it amounts to having two two-component spinors that do not interact with each other. Hence the degrees of freedom are doubled.

Note that the boundary conditions in Eq. (5.22) can only be satisfied if ϕ_k changes sign in going from $x = -\infty$ to $+\infty$. So the topological nature of the kink is essential to the existence of the zero mode.

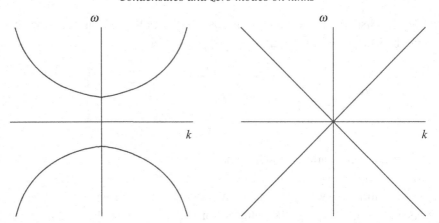

Figure 5.1   The dispersion curve for fermions in the vacuum is shown on the left and for fermion zero modes on the domain wall on the right.

In constructing the zero mode, we have postulated that ξ be an eigenstate of $i\gamma^x$. Therefore there is a possibility that there might be other zero mode solutions. However, it is possible to prove that this is not the case and the zero mode(s) that we have found are the only such solutions. The proof proceeds by choosing a set of orthogonal basis spinors

$$\chi_1 = \begin{pmatrix} 1 \\ 0 \\ 0 \\ i \end{pmatrix}, \quad \chi_2 = \begin{pmatrix} 0 \\ 1 \\ i \\ 0 \end{pmatrix}, \quad \chi_3 = \begin{pmatrix} i \\ 0 \\ 0 \\ 1 \end{pmatrix}, \quad \chi_4 = \begin{pmatrix} 0 \\ i \\ 1 \\ 0 \end{pmatrix} \tag{5.34}$$

The first two spinors are eigenstates of $i\gamma^x$ with eigenvalue -1. These are the spinors that occur in the general solution we have already found subject to the condition that $i\gamma^x \xi = -\xi$. Since the Dirac equation is linear, any new solution must be a linear combination of χ_3 and χ_4. However, both these basis spinors are eigenstates of $i\gamma^x$ with eigenvalue $+1$ and we have seen that such states cannot be part of the solution since the boundary conditions cannot be met. Therefore there are no other zero mode solutions beside the ones that we have already constructed.

As mentioned in the introduction to this chapter, the interpretation of fermionic zero modes is quite distinct from that of bosonic condensates. Fermionic modes should be viewed as states in which the fermions can reside. A mode by itself does not carry energy density or charge or some other physical quantity. Only if the mode is occupied, can it contribute to the energy. However, the zero mode is special in some ways since, even if it is occupied, the fermion occupying the zero mode contributes zero energy. Likewise, if the zero mode is unoccupied, it also contributes zero energy and so the system has a degenerate ground state. Indeed the occurrence of a zero mode leads to some novel and important quantum field theoretic consequences that we shall outline in Section 5.3.

In the discussion of fermion zero modes above we have considered only a Yukawa interaction between the fermion field and the field that makes up the domain wall. More generally, there can also be Majorana interactions. Zero modes of Majorana fermions on domain walls have been discussed in [147].

Just like scalar field condensates and fermionic zero modes on domain walls, there can also be gauge field (or spin-1) condensates. These arise when the model has broken gauge symmetries in addition to broken discrete symmetries. This is precisely the situation in the $SU(5) \times Z_2$ model discussed in Chapter 2 and the kinks in the model have condensates of spin-1 fields as we describe in Section 5.5.

Finally we close this section by remarking that there are several mathematical "index theorems" that can be used to obtain information on the number of zero modes on a soliton [176]. In the case of domain walls that we have been discussing, however, the index theorems do not lead to a useful result.

5.3 Fractional quantum numbers

To quantize a fermionic field we find all the modes (solutions of the Dirac equation) and then associate creation and annhilation operators with each of the modes.[2] The same procedure may be followed in the presence of zero modes [83]. As discussed in the previous section, there is a single zero mode on the Z_2 kink (in one spatial dimension), which is denoted by ψ_0. Then the expansion of the field operator in modes is

$$\psi = a_0 \psi_0 + \sum_p \left[b_p \psi_{p+} + d_p^\dagger \psi_{p-}^c \right] \tag{5.35}$$

The second term is the usual sum over the positive energy modes, ψ_{p+}, and fermion-number conjugates of the negative energy modes, ψ_{p-}^c.[3] There is no sum over spin because there is no spin degree of freedom in one spatial dimension. The first term in Eq. (5.35) contains the zero mode, ψ_0, and a_0 is the operator associated with this mode. The term may seem strange because the zero mode does not have a corresponding conjugated term. This is because $\psi_0^c = \psi_0$ and so the mode functions associated with a_0 and its conjugate operator are identical. However, one still has the usual equal time anticommutation relations for the field and its canonical momentum

$$\{\psi_a(\mathbf{x}), \psi_b^\dagger(\mathbf{y})\} = \delta(\mathbf{x} - \mathbf{y})\delta_{ab} \tag{5.36}$$

---

[2] We work in one spatial dimension in this section and hence spinors have two components.
[3] That is, ψ_{p-}^c is the wavefunction of a hole in the Dirac sea that has momentum p.

while other anticommutators vanish. Using the expansion in terms of creation and annihilation operators this gives

$$\{a_0, a_0\} = \{a_0^\dagger, a_0^\dagger\} = 0, \qquad \{a_0, a_0^\dagger\} = 1 \tag{5.37}$$

Since the Dirac Lagrangian in Eq. (5.14) is invariant if the fermion fields are rotated by a phase, the model has a conserved fermion number current. The Noether current is given by $\bar{\psi}\gamma^\mu\psi$ ($\mu = 0, 1$). In the quantum theory the physical current operator needs to be normal ordered. This is equivalent to defining the fermion number operator as

$$Q_f = \int dx : j^0 := \frac{1}{2}\int dx(\psi_\alpha^\dagger\psi_\alpha - \psi_\alpha\psi_\alpha^\dagger) \tag{5.38}$$

We can act by this operator on any state to determine the fermion number of that state. Let us denote the state with no positive energy particles and empty zero mode by $|0; -\rangle$ and the state with no positive energy particles and filled zero mode by $|0; +\rangle$. Then the fermion numbers of these two states are

$$\begin{aligned} Q_f|0; \pm\rangle &= \frac{1}{2}[a_0^\dagger a_0 - a_0 a_0^\dagger]|0; \pm\rangle \\ &= \frac{1}{2}[2a_0^\dagger a_0 - 1]|0; \pm\rangle \\ &= \pm\frac{1}{2}|0; \pm\rangle \end{aligned} \tag{5.39}$$

Therefore the kink carries a half-integer fermion number of either sign. If the fermion carries electric charge, the electric charge on the kink is also half-integral.

It is critical to not think of the kink as being "kink *plus* fermion." Instead the kink is made of both the bosonic and fermionic fields. Then there are only two states for the kink: one with filled zero mode and the second with the zero mode empty.

Surprising as the half-integer fermion number is, further work in [150, 68] obtained different fractional charges in other systems (see Section 9.1). Indeed, [68] shows that the charges can even be irrational.

5.4 Other consequences

If the bosonic condensate is electrically charged, the domain wall becomes superconducting. To see this in some more detail, consider the case of a complex, electrically charged, scalar field, χ, interacting with the field ϕ that forms a domain wall

$$L = L[\phi] + L[A_\mu] + \frac{1}{2}|D_\mu\chi|^2 - \frac{m_\chi^2}{2}|\chi|^2 - \frac{\lambda_\chi}{4}|\chi|^4 - \frac{\sigma}{2}\phi^2|\chi|^2 \tag{5.40}$$

The first term is the Lagrangian for the Z_2 model and the second is the usual Maxwell Lagrangian for the gauge field A_μ. The covariant derivative is defined by

$$D_\mu = \partial_\mu - iq A_\mu \tag{5.41}$$

The propagating modes of the condensate are

$$\chi = \chi_0(x)e^{i(\omega t - \mathbf{k} \cdot \mathbf{x})} \tag{5.42}$$

where $\chi_0(x)$ is the condensate profile and $\mathbf{k}$ is the wave-vector restricted to lie in the plane of the wall, the yz-plane. Since χ carries electric charge, q, the electric current is

$$\mathbf{j}_\chi = \frac{iq}{2}(\chi^\dagger \nabla \chi - \chi \nabla \chi^\dagger) \tag{5.43}$$

Inserting Eq. (5.42) into (5.43), we find that the current is along the $\mathbf{k}$ direction

$$\mathbf{j}_\chi = q|\chi_0|^2 \mathbf{k} \tag{5.44}$$

The simplest way to see that the wall with the condensate is superconducting is to write

$$\chi = \chi_0(x)e^{i\theta} \tag{5.45}$$

where χ_0 is the condensate solution and θ is the phase variable. Then the low energy Lagrangian for the θ degree of freedom can be derived by integrating the full Lagrangian density, Eq. (5.40), over x to get

$$L(\theta) = \frac{1}{2}(\partial_\mu \theta - eA_\mu)^2 \tag{5.46}$$

where we have omitted an overall constant factor. This effective Lagrangian is the relativistic generalization of the Lagrangian in the Ginzburg-Landau theory of superconductivity. Assuming that the relativistic generalization does not make any qualitative difference, results from the Ginzburg-Landau theory can then be applied directly to the present case. In particular, the domain wall with charged condensate can be expected to carry persistent electric currents, have magnetic vortices, and exhibit the Meissner effect (expulsion of magnetic fields) [61].

We now discuss fermionic superconductivity on domain walls. The relevant modes are given in Eq. (5.33) and the (normal ordered) current is

$$j_\psi = q : \psi^\dagger \gamma \psi : \tag{5.47}$$

Using the expansion of ψ in terms of creation and annihilation operators (Eq. (5.35)), the current in any Fock state of fermions can be evaluated. Similarly, the electric charge on a domain wall can also be evaluated.

Fermions on domain walls can only make the wall superconducting if they form Cooper pairs and condense. It is believed that the slightest attractive interaction between the fermions on the wall will lead to condensation below some critical temperature. On a domain wall, there are possible channels for attractive interactions. For example, the fermions interact with each other via exchange of ϕ quanta and this can lead to an attractive force. The problem of rigorously showing fermionic superconductivity of domain walls has not been investigated. In particular, the Meissner effect, which is the hallmark of superconductivity, has not been shown. Indeed, the response of non-interacting fermion zero modes to an external magnetic field has been discussed with the conclusion that the walls are diamagnetic [173] (also see [82, 172]).

In the particle-physics/cosmology literature, the existence of electrically charged zero modes is simply assumed to imply superconductivity (though see [15]). A reason for this assumption is that a current on a wall persists even without the application of an external electric field. Once the current carrying fermionic zero mode states have been populated there are very few processes by which these states can be emptied [184]. Two such dissipative processes are the scattering of counter-propagating fermion zero modes, and the scattering of fermion zero modes with fluctuations of the domain wall field itself. Generally these processes occur at a very slow rate, at least in astrophysical situations of interest. Hence, strictly speaking, domain walls in particle physics/cosmology have only been shown to be excellent conductors and not superconductors.

The equilibrium current on a domain wall in any setting depends on the balance of the rates of current increase owing to an external electric field and decrease owing to dissipation. Note that an external magnetic field in which a domain wall is moving is, effectively, an electric field in the rest frame of the wall. Since magnetic fields are ubiquitous in astrophysics, any cosmological domain walls with fermion zero modes can be expected to be current carrying. Superconducting domain walls in realistic grand unification models have been discussed in [98].

The fermion zero mode states that we have discussed above are single particle eigenstates. The true states of the domain wall are also affected by fermion-fermion interactions. The many-body problem falls in the class of two-dimensional systems of interacting fermions. In the presence of a strong external magnetic field, so that the Landau level spacing is large compared to other energy scales, the fermions on the wall are similar to electrons in a quantum Hall system.

5.5 Condensates on $SU(5) \times Z_2$ kinks

In Chapter 2 we have discussed kinks in an $SU(5) \times Z_2$ model, which is the simplest example of a Grand Unified Theory. Even though $SU(5)$ grand unification is known not to be phenomenologically viable, the model is still pedagogically useful.

The Lagrangian for the model is

$$L = L_b[\Phi, \chi, X_\mu] + L_f[\chi, \psi, X_\mu] \tag{5.48}$$

where the $SU(5)$ adjoint field, Φ, does not couple directly to the fermionic fields (denoted generically by ψ). Only an additional $SU(5)$ fundamental field, χ, couples to the fermions. The vacuum expectation value of χ is responsible for electroweak symmetry breaking and the masses of all the observed quarks and leptons arise from this symmetry breaking. The $SU(5)$ symmetry breaking has no consequences for the fermionic sector, except via the χ field. This indirect effect can be important in the presence of kinks, since χ can form a condensate on the kink, which can then interact with some of the fermions. We will discuss this further below.

The bosonic part of the Lagrangian is

$$L_b = \text{Tr}(D_\mu \Phi)^2 + |D_\mu \chi|^2 - V(\Phi, \chi) - \frac{1}{2}\text{Tr}(X_{\mu\nu} X^{\mu\nu}) \tag{5.49}$$

The covariant derivative is defined by $D_\mu \equiv \partial_\mu - ig X_\mu$ and its action on the scalar fields is

$$D_\mu \Phi \equiv \partial_\mu \Phi - ig[X_\mu, \Phi], \quad D_\mu \chi \equiv (\partial_\mu - ig X_\mu)\chi \tag{5.50}$$

The potential is given by

$$V(\Phi, \chi) = V(\Phi) + V(\chi) + \lambda_4 (\text{Tr}\Phi^2)\chi^\dagger \chi + \lambda_5(\chi^\dagger \Phi^2 \chi) \tag{5.51}$$

with

$$V(\Phi) = -m_1^2(\text{Tr}\Phi^2) + \lambda_1(\text{Tr}\Phi^2)^2 + \lambda_2(\text{Tr}\Phi^4) \tag{5.52}$$
$$V(\chi) = -m_2^2 \chi^\dagger \chi + \lambda_3(\chi^\dagger \chi)^2 \tag{5.53}$$

Successful grand unification requires that the global minimum of the potential leaves an $SU(3) \times U(1)$ symmetry unbroken.[4] As already described in Section 2.2, the minimum of the potential with χ set equal to zero, occurs at

$$\Phi_0 = \frac{\eta}{2\sqrt{15}}\text{diag}(2, 2, 2, -3, -3) \tag{5.54}$$

(up to $SU(5) \times Z_2$ transformations) in the parameter range

$$\lambda \geq 0, \quad \lambda' \equiv h + \frac{7}{30}\lambda \geq 0 \tag{5.55}$$

The vacuum expectation value of Φ breaks $SU(5) \times Z_2$ to $[SU(3) \times SU(2) \times U(1)]/Z_6$. If we assume that the back-reaction of a vacuum expectation value of χ

---

[4] Symmetry breaking patterns have been discussed quite generally in [99].

on that of Φ is small, we can write down a reduced potential for χ alone

$$V_{\text{red}}(\chi; \Phi_0) = \left(-m_2^2 + \lambda_4 \text{Tr}\Phi_0^2 + \frac{\lambda_5}{15}\eta^2 \right) \chi_a^\dagger \chi_a$$
$$+ \left(-m_2^2 + \lambda_4 \text{Tr}\Phi_0^2 + \frac{3\lambda_5}{20}\eta^2 \right) \chi_b^\dagger \chi_b + \lambda_3(\chi^\dagger \chi)^2 \quad (5.56)$$

where $a = 1, 2, 3$ and $b = 4, 5$. The symmetry is broken to $[SU(3) \times U(1)]/Z_3$ only if the vacuum expectation value of χ is along the χ_4 or χ_5 directions. This further restricts the range of parameters to

$$\frac{\lambda_5}{15}\eta^2 > m_2^2 - \lambda_4 \text{Tr}\Phi_0^2 > \frac{3\lambda_5}{20}\eta^2, \qquad \lambda_3 > 0 \quad (5.57)$$

which also implies $\lambda_5 < 0$ and $m^2 < \lambda_4 \text{Tr}\Phi_0^2$. We assume that these conditions on the parameters are satisfied. Then a minimum of the reduced potential occurs at

$$\chi^T = \eta_{\text{ew}}(0, 0, 0, 1, 0) \quad (5.58)$$

where

$$\eta_{\text{ew}}^2 = \frac{1}{2\lambda_3} \left(m_2^2 - \lambda_4 \text{Tr}\Phi_0^2 - \frac{3\lambda_5}{20}\eta^2 \right) \quad (5.59)$$

is the electroweak symmetry breaking scale. The final $[SU(3) \times U(1)]/Z_3$ symmetry corresponds to the color and electromagnetic symmetries present today.

Next we describe the fermionic sector.[5] There are two fermion fields: ψ, which is in the fundamental (5-dimensional) representation of $SU(5)$ and ζ, which is in the antisymmetric 10-dimensional representation. The known quarks and leptons fit within the components of these fields. With the choice of vacuum expectation values in Eq. (5.58)

$$(\psi^i)_{\text{L}} = (d^{c1}, d^{c2}, d^{c3}, e^-, -\nu_e)_{\text{L}} \quad (5.60)$$

$$(\psi^i)_{\text{R}} = (d_1, d_2, d_3, e^+, -\nu_e^c)_{\text{R}} \quad (5.61)$$

$$(\zeta_{ij})_{\text{L}} = \frac{1}{\sqrt{2}} \begin{pmatrix} 0 & u^{c3} & -u^{c2} & u_1 & d_1 \\ -u^{c3} & 0 & u^{c1} & u_2 & d_2 \\ u^{c2} & -u^{c1} & 0 & u_3 & d_3 \\ -u_1 & -u_2 & -u_3 & 0 & e^+ \\ -d_1 & -d_2 & -d_3 & -e^+ & 0 \end{pmatrix}_L \quad (5.62)$$

(see Eq. (14.9) in [30]). The numerical index on the u and d fields refers to color charge, and the placement (subscript or superscript) depends on the representation

---

[5] Actually we describe only one of the three families of the standard model fermionic sector, and then too the neutrino is taken to be massless.

(fundamental or fundamental conjugate) in which the field transforms under the unbroken $SU(3)$. The c superscript denotes charge conjugation:

$$\psi^c \equiv i\gamma^2 \psi^* \tag{5.63}$$

The L and R subscripts refer to left- and right-handed spinors

$$\psi_L \equiv \frac{1 - \gamma^5}{2}\psi, \qquad \psi_R \equiv \frac{1 + \gamma^5}{2}\psi \tag{5.64}$$

The Dirac γ matrices are defined in Eqs. (5.15) and (5.17).

Now we are ready to describe the interactions of the various fields with the $SU(5) \times Z_2$ kink, described as the $q = 2$ kink in Chapter 2.

- In the presence of a $(q = 2)$ kink, the vacuum expectation values are

$$\Phi(-\infty) = +\frac{\eta}{2\sqrt{15}}\text{diag}(2, 2, 2, -3, -3)$$

$$\chi^T(-\infty) = \eta_{ew}(0, 0, 0, 1, 0)$$

$$\Phi(+\infty) = -\frac{\eta}{2\sqrt{15}}\text{diag}(2, -3, -3, 2, 2)$$

$$\chi^T(+\infty) = \eta_{ew}(0, 0, 1, 0, 0)$$

Note that the non-trivial entry of χ has to coincide with one of the -3 entries of Φ since this is what minimizes the potential $V(\Phi, \chi)$. Therefore χ must rotate through the kink. Inside the kink, the fields are not pure rotations of the asymptotic values.

- The component Φ_{11} goes from $+2$ to -2 as the wall is crossed. Hence it must vanish in the wall. This is very similar to the Z_2 case, where the field vanishes at the center of the wall. The field χ interacts with Φ as given by the potential in Eq. (5.51). Note the interaction term $\lambda_5 \text{Tr}(\chi^\dagger \Phi^2 \chi)$, which directly couples χ_1 to Φ_{11}. (The other term couples all components of χ to $\text{Tr}\Phi^2$ only.) By explicit construction it can be seen that χ_1 can condense on the wall for a certain range of parameters [146]. Hence the $SU(5) \times Z_2$ model allows for scalar condensates on the wall (see Section 5.1). In addition, since χ_1 is a complex scalar field, the condensate has an associated phase. The choice of phase on different parts of the wall may be different, leading to vortices in χ_1 that can only exist on the wall. Since χ_1 transforms non-trivially under the unbroken $SU(3)$, the vortices carry color magnetic field. This is similar to our discussion below Eq. (5.46).

- Next we consider fermion interactions with the wall [146]. The fermions do not couple directly to Φ. Hence the only coupling to the wall is due to the rotation of χ in passage through the wall and to the condensate in the χ_1 component. Consider the scattering of the fifth component, ψ_5, which corresponds to a neutrino on the left side of the wall but a d-quark on the right. This fifth component has non-zero reflection and transmission probability. If it reflects, the particle is still a neutrino. If it transmits, it must change into a d-quark. If a neutrino becomes a d-quark in passing through the wall, it must pick up electric and color charge from the wall. Hence we are forced to conclude that there must be electric and color excitations that live entirely on the wall. If a χ_1 condensate is not present, the only available excitations are the charged gauge field components. Hence charged gauge fields must condense on the wall.

To see the presence of a charged gauge field condensate, it is most convenient to go to a gauge where the scalar field vacuum expectation values are oriented in the same directions on both sides of the wall, as we now discuss.

• Consider a very thin wall, so that

$$\Phi(x < 0) = +\frac{\eta}{2\sqrt{15}}\text{diag}(2, 2, 2, -3, -3) \equiv \Phi_0$$

$$\Phi(x > 0) = -\frac{\eta}{2\sqrt{15}}\text{diag}(2, -3, -3, 2, 2) \tag{5.65}$$

Now we perform a local gauge transformation that rotates Φ into the direction of Φ_0 (up to a sign) everywhere. Such a gauge rotation is local since it is equal to the identity transformation for $x < 0$ but is non-trivial for $x > 0$ since it exchanges the 23- and 45-blocks of Φ. In both regions, $x < 0$ and $x > 0$, the gauge rotation is constant. The rotation is non-constant only at $x = 0$ i.e. on the wall. Hence the gauge fields after the rotation vanish everywhere except on the wall itself and there are gauge degrees of freedom residing on the wall. A more explicit calculation shows that the gauge fields living on the wall carry electric and color charge.

5.6 Possibility of fermion bound states

In addition to fermionic zero modes on a kink, there may also be fermionic bound states. Such bound states would have a non-vanishing energy eigenvalue ω with $0 < \omega < m$. Since the energy eigenvalue is less than the asymptotic mass, the fermion would be bound to the wall. We examine whether the model in Eq. (5.14) leads to fermionic bound states on a Z_2 kink.

For convenience we work in one spatial dimension. Then spinors have two components and there are only two gamma matrices, which can be taken to be

$$\gamma^0 = \sigma^3 = \begin{pmatrix} 1 & 0 \\ 0 & -1 \end{pmatrix}, \qquad \gamma^1 = i\sigma^1 = \begin{pmatrix} 0 & 1 \\ 1 & 0 \end{pmatrix} \tag{5.66}$$

Then the Dirac equation $i\partial\!\!\!/\psi - g\phi_k\psi = 0$ together with $\psi = \exp(-i\omega t)\xi$ gives

$$\partial_x\xi_1 = -(\omega + g\phi_k)\xi_2$$

$$\partial_x\xi_2 = +(\omega - g\phi_k)\xi_1 \tag{5.67}$$

where

$$\xi = \begin{pmatrix} \xi_1(x) \\ \xi_2(x) \end{pmatrix} \tag{5.68}$$

and we are interested in solutions with

$$0 < \omega < m_f \equiv g\eta \tag{5.69}$$

The boundary conditions at the origin for ξ_1 and ξ_2 may be determined by noting that we are free to rescale both ξ_1 and ξ_2 by a constant factor. So we can set $\xi_1(0) = +1$. Further, using the symmetry $\phi_k(x) = -\phi_k(-x)$, we find that the equations are invariant if we replace $\xi_1(x)$ by $c\xi_2(-x)$ and $\xi_2(x)$ by $c\xi_1(-x)$, where c is a constant. Hence

$$\xi_1(x) = c\xi_2(-x), \qquad \xi_2(x) = c\xi_1(-x) \tag{5.70}$$

This gives

$$\xi_1(x) = c\xi_2(-x) = c^2\xi_1(x) \tag{5.71}$$

Since $\xi_1(x)$ cannot vanish for all x, we get

$$c = \pm 1 \tag{5.72}$$

Therefore there are two possible boundary conditions at the origin

$$\xi_2(0) = \pm\xi_1(0) = \pm 1 \tag{5.73}$$

At infinity we require $\xi_1 \to 0$ and $\xi_2 \to 0$.

A numerical search for a solution with non-zero ω did not reveal any bound states for the range of parameters $0.1 < m_f w < 20$, where w is the width of the kink. However this does not exclude the existence of fermion bound states (beside the zero mode) on kinks in other systems, and it remains an open problem to find systems where such bound states exist.

5.7 Open questions

1. Explore the classical and quantum physics of a domain wall with electrically charged bosonic and fermionic zero modes placed in an external magnetic field. What happens if the domain wall is moving?
2. Calculate the reflection of photons off a superconducting domain wall. Is the wall a good mirror? (See [184].)
3. Construct a system in which the kink has both a zero mode and a fermionic bound state.

6

Formation of kinks

In this chapter we study the formation of kinks and domain walls during a phase transition. We start by describing the effective potential for a field theory at finite temperature. This sets up a useful framework for discussing phase transitions and defect formation.

6.1 Effective potential

The effective potential is a tool that is often used to study phase transitions in field theory [89, 179, 90, 47, 100]. The idea is to consider the interaction of a scalar degree of freedom ("order parameter") with a thermal background of particles. Such processes induce additional temperature dependent terms in the potential for the order parameter, leading to an "effective potential." The shape of the effective potential varies as a function of temperature and new minima might appear. The global minimum defines the vacuum of the model. If a new global minimum appears at some temperature, it indicates that the system makes a transition to a new expectation value of the order parameter and there is a phase change. We now describe the (one loop) effective potential in a little more detail.

We consider a field theory of scalar, spinor and vector fields

$$L = L_B + L_F \tag{6.1}$$

with the bosonic Lagrangian

$$L_B = \frac{1}{2}(D_\mu \Phi_i)D^\mu \Phi_i - V(\Phi) - \frac{1}{4}F^a_{\mu\nu}F^{\mu\nu a} \tag{6.2}$$

where Φ_i are the components of the scalar fields,

$$D_\mu \equiv \partial_\mu - ieA^a_\mu T^a \tag{6.3}$$

the T^a are group generators, and

$$F_{\mu\nu}^a \equiv \partial_\mu A_\nu^a - \partial_\nu A_\mu^a + e f^{abc} A_\mu^b A_\nu^c \tag{6.4}$$

where A_μ^a are the gauge fields.

The Lagrangian for a fermionic multiplet Ψ is

$$L_F = i\bar{\Psi}\gamma^\mu D_\mu \Psi - \bar{\Psi}\Gamma_i \Psi \Phi_i \tag{6.5}$$

where Γ_i are the Yukawa coupling matrices. The quantity Ψ denotes a collection of fermionic fields and the Yukawa coupling term may be written more explicitly as $\bar{\Psi}_\alpha^\sigma \Gamma_{i\sigma\rho}^{\alpha\beta} \Psi_\beta^\rho \Phi_i$ where α, β label the various fermionic fields, the superscripts σ, ρ on the fermion fields are spinor indices, and i labels the interaction term with the scalar field Φ_i. Γ_i has spinor indices because it could contain the unit matrix (vector coupling) and/or the γ^5 matrix (axial coupling) defined in Eq. (5.17).

If the expectation values of the scalar fields are denoted by Φ_{0i}, then the mass matrices of the various fields are written as

$$\mu_{ij}^2 = \left. \frac{\partial^2 V}{\partial \Phi_i \partial \Phi_j} \right|_{\Phi=\Phi_0}, \quad \text{scalar fields} \tag{6.6}$$

$$m = \Gamma_i \Phi_{0i}, \quad \text{spinor fields} \tag{6.7}$$

$$M_{ab}^2 = e^2 (T_a T_b)_{ij} \Phi_{0i} \Phi_{0j}, \quad \text{vector fields} \tag{6.8}$$

where a, b are gauge field group indices.

Then the finite temperature, one loop effective potential is[1]

$$V_{\text{eff}}(\Phi_0, T) = V(\Phi_0) + \frac{\mathcal{M}^2}{24} T^2 - \frac{\pi^2}{90} \mathcal{N} T^4 \tag{6.9}$$

where

$$\mathcal{N} = \mathcal{N}_B + \frac{7}{8} \mathcal{N}_F \tag{6.10}$$

is the number of bosonic and fermionic spin states, and

$$\mathcal{M}^2 = \text{Tr}(\mu^2) + 3\text{Tr}(M^2) + \frac{1}{2}\text{Tr}(\gamma^0 m \gamma^0 m) \tag{6.11}$$

where γ^0 is defined in Eq. (5.15). Note that $\mathcal{M}^2$ depends on the expectation value Φ_0 through the defining equations for the mass matrices given above. For example, $\mathcal{M}^2$ contains a term proportional to $\text{Tr}(\Phi_0^2)$.

An important feature of the effective potential is that it can show the presence of phase transitions. If there are scalar fields with negative mass squared terms in

---

[1] Radiative corrections and spontaneous symmetry breaking are discussed in [33, 178].

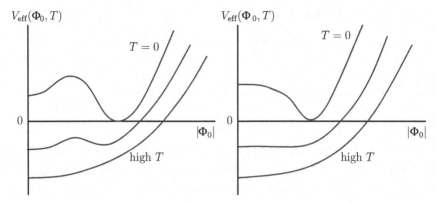

Figure 6.1   Sketch of effective potential for first-order phase transition (left) and second-order phase transition (right). In the first-order case, the global minimum of the potential at high temperature ($\Phi_0 = 0$ in illustration) becomes a local minimum at low temperature. In the second-order case, the global minimum of the potential at high temperature becomes a local maximum at low temperature. The effective potential at $\Phi_0 = 0$ decreases with increasing temperature because of the last term proportional to $-\mathcal{N}T^4$ in Eq. (6.9).

$V(\Phi)$, the contributions from the $\mathcal{M}^2 T^2$ term in the effective potential, Eq. (6.9), can make the effective mass squared positive for these fields if the temperature is high enough (see Fig. 6.1). Therefore when the system is at high temperature, the effective mass squared can be positive and the minimum of the potential at $\Phi_0 = 0$. As the system is cooled, the effective mass squared becomes negative and the minimum of the effective potential occurs at non-zero values of Φ_0 and the lowest energy state has shifted from $\Phi_0 = 0$ to $\Phi_0 \neq 0$. The order parameter, Φ, acquires a non-zero "vacuum expectation value" at some critical temperature. This is the phenomenon of spontaneous symmetry breaking and manifests itself as a phase transition. The phase at high temperature had a certain symmetry dictated by the invariance of the field theory with $\Phi_0 = 0$ and at low temperature the symmetry is changed because now $\Phi_0 \neq 0$.

As a simple example of an effective potential, consider the $\lambda\phi^4$ model of Eq. (1.2) with

$$V(\phi) = -\frac{m_0^2}{2}\phi^2 + \frac{\lambda}{4}\phi^4 + \frac{\lambda}{4}\eta^4 \qquad (6.12)$$

Then μ^2 of Eq. (6.6) is given by

$$\mu^2 = -m_0^2 + 3\lambda\phi_0^2 \qquad (6.13)$$

and since there is only one scalar field in this model

$$\mathcal{M}^2 = -m_0^2 + 3\lambda\phi_0^2 \qquad (6.14)$$

Therefore, up to a term that is independent of ϕ_0, the effective potential becomes

$$V_{\text{eff}}(\phi_0, T) = \frac{\bar{m}^2}{2}\phi_0^2 + \frac{\lambda}{4}\phi_0^4 \tag{6.15}$$

with

$$\bar{m}^2 = -m_0^2 + \frac{\lambda}{4}T^2 \tag{6.16}$$

Note that the masses of small excitations around the true vacuum are given by $V_{\text{eff}}''(\phi_0)$ with ϕ_0 being the vacuum expectation value. By minimizing V_{eff} (Eq. (6.15)) we get

$$\phi_{0,\text{min}}(T) = 0, \quad \bar{m}^2 > 0 \tag{6.17}$$

$$= \sqrt{\frac{-\bar{m}^2}{\lambda}}, \quad \bar{m}^2 < 0 \tag{6.18}$$

leading to the mass squared for small excitations (particles) in the true vacuum

$$m_{\text{eff}}^2 \equiv V_{\text{eff}}''(\phi_{0,\text{min}}) = \frac{\lambda}{4}(T^2 - T_c^2), \quad T > T_c \tag{6.19}$$

$$= \frac{\lambda}{2}(T_c^2 - T^2), \quad T < T_c \tag{6.20}$$

where T_c is the critical temperature

$$T_c = \frac{2m_0}{\sqrt{\lambda}} \tag{6.21}$$

In cosmology, since the universe is expanding, it is also cooling. Therefore we can have one or many cosmological phase transitions and the particle-physics symmetries at high temperatures (early universe) and low temperatures (recent universe) are different. The symmetry after the phase transition can be smaller or larger than the symmetry before the phase transition. In other words, lowering the temperature can spontaneously break or restore a symmetry. We will mostly consider symmetry breaking during the phase transition but examples of symmetry restoration are also easy to construct. A system in which symmetry restoration is observed is Rochelle salt [85, 179].

6.2 Phase dynamics

The effective potential $V_{\text{eff}}(\Phi_0, T)$ is calculated for a system that is in thermal equilibrium, assuming a homogeneous vacuum expectation value of the order parameter Φ_0. Yet thermal equilibrium is not maintained during the phase transition and also the phase change occurs in an inhomogeneous manner. The dynamics are clear

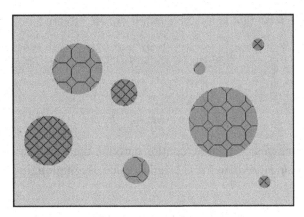

Figure 6.2   A schematic diagram of bubbles nucleating in a first-order phase transition. The two types of bubbles correspond to the two different values of the order parameter. The bubbles grow and collide, and new bubbles nucleate as well. Eventually the whole system is in the new phase.

for a first-order phase transition in which the high temperature phase becomes a metastable state (see Fig. 6.1) at some critical temperature. Now the system can be stuck in this metastable state even when the temperature drops significantly below the critical temperature. An external perturbation can cause the system to transition to the global vacuum. In the absence of an external perturbation, quantum tunneling can trigger the transition. In either case, bubbles of a critical size of the true vacua ($\Phi_0 \neq 0$) nucleate in the false vacuum ($\Phi_0 = 0$) background (see Fig. 6.2). These bubbles grow and eventually merge thus filling space and completing the phase transition. Clearly this process is not homogeneous and cannot be described by an effective potential.

In a second-order transition, in contrast to a first-order transition, there is no metastable state in which the system can be trapped. Thus Φ_0 evolves continuously ("spinodal decomposition") from $\Phi_0 = 0$ to $\Phi_0 \neq 0$. However, different spatial regions evolve at different rates owing to thermal and quantum fluctuations, and Φ_0 is not spatially uniform. Once again, since the effective potential assumes constant Φ_0, it can indicate the existence of a second-order phase transition but cannot be expected to accurately describe the dynamics of the transition. Since defect formation crucially depends on the inhomogeneities of the order parameter during the transition, new ideas have been needed to predict the statistical properties of defects formed in a second-order phase transition.

In one spatial dimension, the distribution of kinks is described by the number density of kinks, and correlators of kink locations. In higher dimensions, the problem becomes richer because domain walls are extended and can curve and have complicated topology. In addition to the mass density in domain walls, we are interested in the statistical distribution of shapes and sizes of domain walls formed at the phase transition.

6.3 Kibble mechanism: first-order phase transition

At a first-order phase transition, the order parameter has to change from $\Phi_0 = 0$ to its non-zero vacuum expectation value. We are interested in the case when there is more than one possible non-zero value for Φ_0. Then the dynamics in a small spatial region select a vacuum. However the vacuum selected in different spatial regions can be different. For example, in the case of the Z_2 model, the field in a certain region might relax into the $\phi = +\eta$ vacuum, whereas in another region it might relax into $\phi = -\eta$ (see Fig. 6.2).

In a first-order phase transition, each bubble is filled with constant Φ_0 i.e. a fixed vacuum is chosen within a bubble but it can be different for different bubbles. With time, the bubbles grow and collide and fill up the volume. Let us denote by ξ the characteristic size of a region where the same vacuum is selected, after the phase transition is over. Then ξ is the typical size of bubbles when they percolate. If Γ denotes the bubble nucleation rate per unit volume and v is the velocity of the growing bubble walls, then we can define a length scale and a time scale on dimensional grounds (in D spatial dimensions) by

$$\tilde{\xi} = \left(\frac{v}{\Gamma}\right)^{1/(D+1)}, \qquad \tilde{\tau} = \left(\frac{1}{v^D \Gamma}\right)^{1/(D+1)} \tag{6.22}$$

The domain size ξ is a numerical factor times $\tilde{\xi}$ and in practice we take $\xi \sim \tilde{\xi}$. Similarly $\tilde{\tau}$ is related to the time that it takes to complete the phase transition.[2]

The process of bubble percolation has been studied both analytically and numerically [95]. Taking the centers of the bubbles as the vertices of a lattice and connecting only the centers of bubbles that collide, we obtain a random lattice (see Fig. 6.3). We would like to determine the characteristics of such a random lattice since this plays a role in determining the network of defects that form. For example, the typical number of bubbles with which any given bubble collides, also known as the "coordination number" of the random lattice, plays a role in the fraction of closed topological defects (closed domain walls, loops of string, or closely paired monopole-antimonopole pair) that are formed.

In one spatial dimension, every bubble trivially collides with two other bubbles. In two spatial dimensions the average number of collisions is the same as the coordination number of a fully triangulated lattice of infinite extent. From purely

---

[2] Equation (6.22) relates ξ to the nucleation rate Γ, but it is very hard to measure Γ in any experiment. In fact, it may be easier to measure properties of the defect network, then ξ, and from it infer Γ.

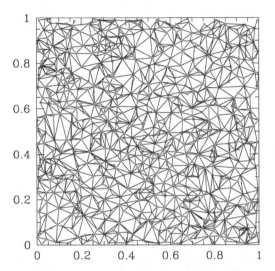

Figure 6.3   If two bubbles collide, their centers are joined by straight lines. The figure then shows the "random bubble lattice" expected in a first-order phase transition in two spatial dimensions.

geometrical constraints that we describe next, the coordination number is six (see, for example, [129]).

The lattice is infinite in extent and by identifying the points at infinity we can view the lattice as lying on a two-dimensional sphere. Then Euler's formula relates the number of vertices (V), edges (E) and faces (F) of the lattice

$$V - E + F = 2 \tag{6.23}$$

Let the coordination number be n. Therefore for every vertex there are n edges but every edge is bounded by two vertices. This relates the number of edges to the number of vertices: $E = nV/2$. Also, every face is a triangle, giving three edges to every face. But an edge is shared by two faces. So $E = 3F/2$. Putting together these relations in Euler's formula gives

$$V - nV/2 + nV/3 = 2 \tag{6.24}$$

In the limit of $V \to \infty$, this yields $n = 6$.

In three spatial dimensions, similar arguments have been given [95] to show that the average coordination number is 13.4. This result is not completely fixed by geometrical constraints as in two dimensions and the result can vary a little depending on the details of the bubble size distribution.

Returning to the $\lambda \phi^4$ model, each bubble either has the phase $\phi_0 = +\eta$ or $\phi_0 = -\eta$ within it. If bubbles of different phases collide, a domain wall forms between the centers of those bubbles. If bubbles of the same phase collide, a wall does not form, though it is possible that a closed domain wall or a wall-antiwall pair forms owing to the energetics of the bubble collision. We expect small closed walls and closely separated walls and antiwalls to annihilate. Hence the distribution of domain walls after the phase transition is simply described by the locations of bubble collisions when the bubbles carry different phases. Since the phase in the bubbles is $\pm\eta$ with equal probability, the phase transition is simulated by assigning $\pm\eta$ to each of the vertices of the random bubble lattice as in Fig. 6.3. We shall further discuss the properties of the wall network at formation during a first-order phase transition in Section 6.6.

As we have seen, a first-order phase transition is relatively simple to conceptualize. A second-order phase transition is harder to understand. To discuss second-order phase transitions, it is useful to first define an equilibrium correlation length.

6.4 Correlation length

The "equilibrium correlation length," $\bar{\xi}$, is defined as the distance over which field correlations are significant. Generally the field correlations at two spatial points fall off exponentially with increasing separation between the points, $\exp(-r/\bar{\xi})$, and the exponent defines the equilibrium correlation length, $\bar{\xi}$. Hence we need to evaluate the correlation function

$$G(r) = \langle T | \phi(t, \mathbf{x})\phi(t, \mathbf{y}) | T \rangle \tag{6.25}$$

where G only depends on $r \equiv |\mathbf{x} - \mathbf{y}|$ because the system is translationally invariant. The thermal state is denoted by $|T\rangle$ and is defined as the state containing the equilibrium number density distribution (Fermi-Dirac or Bose-Einstein) of particles

$$|T\rangle = |\{n_{\mathbf{k}}\}\rangle_T \tag{6.26}$$

$$n_{\mathbf{k}} = \frac{1}{e^{\beta\omega_k} \pm 1} \tag{6.27}$$

where $\beta \equiv 1/T$ and ω_k is the energy of particles in the $\mathbf{k}$ mode.[3]

With the Z_2 model in mind, we have only one scalar field and the quantum field operator can be expanded in modes about the true vacuum

$$\phi(t, \mathbf{x}) = \phi_0(T) + \int \frac{d^3 k}{(2\pi)^3} \frac{1}{\sqrt{2\omega_k}} [e^{-i\omega_k t + i\mathbf{k}\cdot\mathbf{x}} a_{\mathbf{k}} + e^{+i\omega_k t - i\mathbf{k}\cdot\mathbf{x}} a_{\mathbf{k}}^\dagger] \tag{6.28}$$

where $\phi_0(T)$ is the vacuum expectation value of the field at temperature T, and $a_{\mathbf{k}}$ and $a_{\mathbf{k}}^\dagger$ are annihilation and creation operators. The dispersion relation is that for a free particle with temperature dependent mass $m(T)$ (see Eq. (6.20) for the $\lambda\phi^4$ model; we have dropped the subscript "eff" for convenience)

$$\omega_k^2 = \mathbf{k}^2 + m^2 \tag{6.29}$$

---

[3] The chemical potential vanishes in the present case.

The thermal state $|T\rangle$ contains a Bose-Einstein distribution of the particle excitations and the number of scalar particles at momentum $\mathbf{k}$ is given by

$$n_{\mathbf{k}} = \frac{1}{e^{\beta \omega_k} - 1} \tag{6.30}$$

where $\beta = 1/T$ (Boltzmann constant has been set to 1).

By inserting the expansion in Eq. (6.28) in the correlator, we find

$$G(r) = \int \frac{d^3k}{(2\pi)^3} \frac{1}{\sqrt{2\omega_k}} \int \frac{d^3p}{(2\pi)^3} \frac{1}{\sqrt{2\omega_p}} e^{i(\mathbf{k}\cdot\mathbf{x}-\mathbf{p}\cdot\mathbf{y})} \langle T|a_{\mathbf{k}}^{\dagger} a_{\mathbf{p}}|T\rangle + K \tag{6.31}$$

where K is a constant which is independent of temperature and proportional to $\delta^{(3)}(\mathbf{x} - \mathbf{y})$. Then

$$G(r) = \int \frac{d^3k}{(2\pi)^3} \frac{e^{-i\mathbf{k}\cdot(\mathbf{x}-\mathbf{y})}}{e^{\beta \omega_k} - 1} + K$$
$$= \frac{T}{4\pi r} e^{-m(T)r} + K \tag{6.32}$$

where, in doing the integral, we have assumed $m(T) \ll T$. From the final expression we get the equilibrium correlation length

$$\bar{\xi} = \frac{1}{m(T)} \tag{6.33}$$

For the Z_2 model (Eq. (6.20))

$$m^2 = \frac{\lambda}{4}(T^2 - T_c^2), \quad T > T_c \tag{6.34}$$

$$= \frac{\lambda}{2}(T_c^2 - T^2), \quad T < T_c \tag{6.35}$$

Therefore the equilibrium correlation length is

$$\bar{\xi}(T) = \sqrt{\frac{4}{\lambda}} \frac{1}{\sqrt{T^2 - T_c^2}}, \quad T > T_c \tag{6.36}$$

$$= \sqrt{\frac{2}{\lambda}} \frac{1}{\sqrt{T_c^2 - T^2}}, \quad T < T_c \tag{6.37}$$

or

$$\bar{\xi}(T) \propto |T - T_c|^{-1/2} \tag{6.38}$$

The essential feature in $\bar{\xi}$ is the singularity at $T = T_c$, that occurs at a time that we denote t_c. Assuming that the cooling (quench) occurs at a constant rate T_c/τ_{ext} (in a range of temperature around T_c), we have

$$T - T_c = -\frac{T_c}{\tau_{\text{ext}}}(t - t_c) \tag{6.39}$$

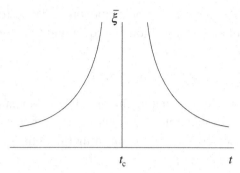

Figure 6.4   Sketch of equilibrium correlation length as a function of time as given in Eq. (6.40).

hence we write

$$\bar{\xi} \sim |T - T_c|^{-\nu} \propto |t - t_c|^{-\nu} \qquad (6.40)$$

for T close to T_c. The exponent ν is called a "critical exponent" and the mean field theory calculation described above gives $\nu = 1/2$. However, the mean field theory ignores particle interactions and renormalization group methods give $\nu = 2/3$, which is closer to experiment. A sketch of the shape of $\bar{\xi}$ (Eq. (6.40)) is shown in Fig. 6.4.

One shortcoming of the mean field calculation of $\bar{\xi}$ is that we have quantized the field ϕ in a fixed true vacuum so that $\phi_0(T)$ in Eq. (6.28) is independent of x. This assumes that the same vacuum is chosen everywhere below T_c. On the other hand, we are precisely interested in the spatial extent of a region in a single vacuum. Hence a more suitable expansion of ϕ would be

$$\phi(t, \mathbf{x}) = \phi_0(t, \mathbf{x}, T) + \int \frac{d^3k}{(2\pi)^3} \frac{1}{\sqrt{2\omega_k}} \left[f_k(t, x)a_{\mathbf{k}} + f_k^*(t, x)a_{\mathbf{k}}^\dagger \right] \qquad (6.41)$$

instead of Eq. (6.28). The vacuum expectation value, ϕ_0, is now allowed to depend on both t and x since the background domain walls may be non-static. The second term in the expansion describes small fluctuations (particles) with mode functions f_k in the classical background $\phi_0(t, x, T)$ at x.

The expansion in Eq. (6.41) is not as obvious as might appear at first sight. We have seen in Chapter 4 that a kink can itself be written in terms of particles via the Mandelstam operator. This was done in the sine-Gordon model but it is conceivable that such an operator also exists in other models. So the above expansion only makes sense for a state in which there is a clear separation between the particle degrees of freedom appearing in the sum and the soliton degrees of freedom included in ϕ_0. For example, if the walls are very close to each other the separation of the two

terms may not be justified. Hence the phenomenon of defect formation is closely
tied to the separation of classical (soliton) and quantum (particle) variables.

We are interested in

$$\xi_0 = \langle T | \phi_0(t, x, T) \phi_0(t, y, T) | T \rangle \tag{6.42}$$

However, we have no way of calculating this equal time "domain correlation function" since (i) the thermal state refers to a thermal distribution of particles, not of
domains, and (ii) the defects do not remain in thermal equilibrium with the particles. This impasse is made less severe by realizing that the calculation of $\bar{\xi}$ for
$T > T_c$ does not suffer from this problem since then $\phi_0 = 0$ is the unique vacuum.
We expect the correlations for $T < T_c$ to be determined by those for $T > T_c$ and so
it might be sufficient to know the correlation length for $T > T_c$. (We discuss this
further in Section 6.5.) To emphasize this point, the two branches of the sketch in
Fig. 6.4 correspond to two different quantities – the equilibrium correlation length
for $t < t_c$ is for excitations in a different vacuum from that for $t > t_c$. So, while
Eq. (6.40) describes the correlations of particle excitations in a given true vacuum
for $T < T_c$ ($t > t_c$), we cannot expect that this has anything to do with the size of
domains of constant vacuum.

The next subtlety is that the divergence of the equilibrium correlation length at
$T = T_c$ should not be taken too literally. The reason is that there is an external
agency (refrigerator) driving the phase transition on a time scale given by τ_{ext}. As
the system gets closer to the critical temperature, it takes longer for equilibrium
to be established, while the external agency continues to cool the system at a rate
determined by external factors. At some temperature above the critical temperature
the time taken to maintain equilibrium becomes larger than the time scale at which
the external conditions are changing.

Assume that the external temperature is being lowered at a constant rate

$$T = T_c \left(1 - \frac{t}{\tau_{\text{ext}}} \right) \tag{6.43}$$

where T_c is the critical temperature and we have chosen $T_c = 0$ for convenience.
The equilibrium correlation length has the form

$$\bar{\xi}(T) = \eta |\epsilon|^{-\nu} \tag{6.44}$$

where ν is a critical exponent, η is some unspecified length scale, and

$$\epsilon \equiv 1 - \frac{T}{T_c} \tag{6.45}$$

The rate of change of $\bar{\xi}$ is

$$\frac{d\bar{\xi}}{dt} = -\frac{\epsilon}{|\epsilon|} \frac{\nu \eta}{\tau_{\text{ext}}} |\epsilon|^{-(\nu+1)} = -\frac{\nu \bar{\xi}}{\tau_{\text{ext}} \epsilon} \tag{6.46}$$

This equation shows that as $\epsilon \to 0-$, $\bar{\xi}$ must change at an ever faster rate if equilibrium is to be maintained.

The relaxation rate can be obtained by perturbing the system and finding how long it takes for the perturbation ("sound") to equilibrate. The result is the "relaxation time"

$$\tau_{\text{rel}} = \tau_0 |\epsilon|^{-\mu} \tag{6.47}$$

where μ is another critical exponent. Then the "speed of relaxation" is the sound speed

$$c_s(T) = \frac{\bar{\xi}}{\tau_{\text{rel}}} = \frac{\eta}{\tau_0} |\epsilon|^{\mu-\nu} \tag{6.48}$$

Note that τ_{rel} diverges as T approaches T_c. This is called "critical slowing down." When the system cannot keep up with the external changes, equilibrium is lost. Denoting the temperature at which τ_{rel} becomes equal to τ_{ext} by T_* we find

$$T_* = T_c \left[1 + \left(\frac{\nu\tau_0}{\tau_{\text{ext}}} \right)^{1/(\mu+1)} \right] \tag{6.49}$$

which occurs at

$$t_* = -\tau_{\text{ext}} \left(\frac{\nu\tau_0}{\tau_{\text{ext}}} \right)^{1/(\mu+1)} \tag{6.50}$$

So we expect the correlation length ξ for $T > T_c$ to be equal to the equilibrium correlation length $\bar{\xi}$ until time t_*, after which ξ departs from $\bar{\xi}$ and grows more slowly (see Fig. 6.5). The behavior of ξ between t_* and t_c is not known and it is generally assumed that ξ does not change very much in this interval. After t_c, there are two distinct vacua, and we need to consider both the correlation scale of chosen vacua (denoted by ξ_0) and the correlations of excitations within a chosen vacuum, ξ. As time goes by, walls annihilate and the domain size with a given vacuum grows. We discuss ξ_0 in the next section.

6.5 Kibble-Zurek mechanism: second-order phase transition

The domain correlation length, ξ_0, over which the same vacuum is chosen, is different from the equilibrium correlation length denoted by $\bar{\xi}$ (Eq. (6.40)). It is also different from the correlation length ξ obtained for particle excitations, including the phenomenon of critical slow down, since ξ_0 has nothing to do with particle excitations. We now discuss different approaches to estimating ξ_0 (for a review see [10]).

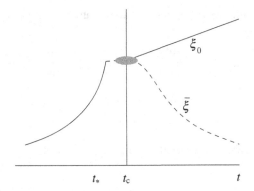

Figure 6.5   The correlation length at high temperature ($t < t_c$) increases as the critical temperature is approached, departing from the equilibrium correlation length when critical slowing down becomes important at t_*. Below the critical temperature, there are two correlation length scales of interest. The domain correlation length, ξ_0, relates to the extent of the spatial domains that are in the same vacuum. This is precisely the spacing of domain walls. Wall-antiwall annihilations cause ξ_0 to grow with time. The particle correlation length, ξ, however, decreases with time since the mass of the particles grows, and eventually approaches the zero temperature value. The dynamics of how ξ_0 separates from $\bar{\xi}$ are not understood and are denoted by the shaded region.

To estimate ξ_0 in the case of a second-order phase transition, Kibble [87] used two different criteria. First, he obtained an upper bound to ξ_0 in the cosmological context based on causality considerations. If the phase transition takes place at a certain cosmic time τ, then the vacua at points separated by more than $c\tau$, where c is the speed of light, must have been selected independently since $c\tau$ is the size of the cosmic horizon. Hence $\xi < c\tau$. This is the "causality bound."

The second estimate is based on finding the Ginzburg length. This is the length over which the choice of vacuum cannot change owing to thermal fluctuations. For concreteness, let us imagine that there is a domain of size l in which $\phi = +\phi_0(T)$ ($T < T_c$) in the Z_2 model. In one spatial dimension this corresponds to a wall-antiwall separated by a distance l, and in three dimensions it corresponds to a closed domain wall of characteristic size l. The idea is that, if l is small, thermal fluctuations can spontaneously change the phase within the domain from $\phi = +\phi_0$ to $\phi = -\phi_0$. However, if l is large, the phase in the domain is frozen, and the distribution of defects does not change spontaneously owing to thermal fluctuations. The smallest length l for which a domain is frozen defines the distance between closest defects and hence predicts the number density of defects.

The energy required to change the phase in a volume R^3 is given by $R^3 \Delta V(T)$ where ΔV is the free energy density difference between the minimum and maximum of the potential at a temperature T. The thermal fluctuation energy available per excitation mode is T according to equipartition. Equating the required and the available energies gives

$$R^3 \Delta V(T) \approx T \tag{6.51}$$

Therefore, at temperature $T < T_c$, a region that is smaller than

$$R \sim \left(\frac{T}{\Delta V(T)}\right)^{1/3} \tag{6.52}$$

will have enough thermal energy to fluctuate from one vacuum to the other. For example, in the Z_2 model (see Eq. (6.15))

$$\Delta V(T) = \frac{\bar{m}^4}{4\lambda} = \frac{\lambda}{64}(T^2 - T_c^2)^2 \tag{6.53}$$

Therefore, at temperature $T < T_c$, the length scale below which regions are still fluctuating are

$$R(T) = \frac{4}{\lambda^{1/3}}\left[\frac{T}{(T_c^2 - T^2)^2}\right]^{1/3} \sim \frac{4^{2/3}}{\lambda^{1/3}T_c}\left[1 - \frac{T}{T_c}\right]^{-2/3} \tag{6.54}$$

where the last approximation holds for $T \sim T_c$.

For a region to fluctuate from one vacuum to another, not only does it need the energy to jump over the barrier, but all different parts of the region need to jump together. This means that all the particles in the domain should be activated coherently. The particle coherence scale is described by the correlation length, which is approximated by the equilibrium correlation length, $\bar{\xi}$. Therefore, at a temperature T, regions of size less than $l_f = \min(R(T), \bar{\xi}(T))$ (subscript "f" stands for "fluctuating") can actively change vacua. The Ginzburg temperature, T_G, is defined by the condition $R(T_G) = \bar{\xi}(T_G)$, and the Ginzburg length is defined by $l_G = \bar{\xi}(T_G)$. For the Z_2 model, this gives

$$T_c - T_G \approx \lambda T_c \tag{6.55}$$

$$l_G = \bar{\xi}(T_G) \approx \frac{1}{\lambda T_c} \tag{6.56}$$

Early estimates took the Ginzburg temperature to be the epoch when domain walls are formed. The number density of walls then follows by dimensional analysis as $\sim 1/l_G^D$.

The relevance of the Ginzburg temperature for defect formation is not clear. As discussed in the previous section, the correlation length $\bar{\xi}$ is calculated for particle excitations in a given vacuum, whereas we are interested in the correlation length of the vacuum domains denoted by ξ_0. In fact, experiments in He-3 find that defects are produced at a temperature below T_c but above T_G, implying that the Ginzburg criterion is not a necessary condition for defect formation. A discussion of the relevance of the Ginzburg criterion in the context of vortex formation in He-3 and He-4 may be found in [86].

Zurek estimated the domain size, ξ_Z, by considering the time scales involved during the phase transition [187, 188, 93]. As discussed at the end of Section 6.4, the system cannot keep up with external changes at $t = t_*$ (Fig. 6.5). Zurek postulated that the correlation length at the instant when the system can no longer keep up with the external changes determines the size of the domains that get frozen. This in turn determines the number of defects.

To estimate ξ_0 at t_c we know $\bar{\xi}$ at the time critical slowing down becomes important. To this we add the distance that a perturbation can propagate from the slow-down time, t_*, to the phase transition time, t_c (see Fig. 6.5). That gives us

$$\xi_0(t_c) = \bar{\xi}(t_*) + \int_{t_*}^{t_c} dt \, c_s(t) \tag{6.57}$$

$$= \eta \frac{1 + \mu}{1 + \mu - \nu} \left(\frac{\tau_{ext}}{\nu \tau_0} \right)^{\nu/(1+\mu)} \tag{6.58}$$

The crucial part of this relation is

$$\xi_0(t_c) \propto \left(\frac{\tau_{ext}}{\tau_0} \right)^{\nu/(1+\mu)} \tag{6.59}$$

This relation gives the dependence of the number density of domain walls in D spatial dimensions on the external time

$$n \propto \left(\frac{\tau_{ext}}{\tau_0} \right)^{-\nu D/(1+\mu)} \tag{6.60}$$

We can control τ_{ext} in experiments and hence this is a testable prediction.

The above analysis can be improved yet further. For example, we have calculated ξ_0 at $t = t_c$. Yet thermal fluctuations after $t = t_c$ (i.e. $T < T_c$) may be important and the domain structure may freeze out at yet lower temperatures, as in the discussion of the Ginzburg length scale above. So the relevant time at which ξ_0 is stable to thermal fluctuations is somewhat after t_c, in agreement with the analysis of [8].

There is yet another view of defect formation at a phase transition first proposed in [5]. In numerical simulations of a $U(1)$ field theory, the authors found that there is a distribution of vortices even at temperatures above the phase transition. However, these vortices are small, closed structures. At the critical temperature, the vortices link up and form infinite, open structures. Thus the phase transition is coincident with a percolation transition of the vortices. If this feature is generally true, we expect a population of small, closed domain walls to exist above the critical temperature. As the temperature is lowered, the walls connect and grow larger and

Table 6.1   *Size distribution of black clusters found by simulations*
on a cubic lattice.

Cluster size	1	2	3	4	6	10	31 082
Number	462	84	14	13	1	1	1

at the critical temperature, the walls percolate, giving walls of infinite extent. The percolation picture has not been checked by simulating the domain wall forming phase transition. However, we can still study the statistical properties of the network of walls formed after a phase transition using some simple arguments that we now describe.

The topic of defect formation and, more generally, phase transition dynamics is still under active investigation.

6.6 Domain wall network formation

The previous sections focused on the density of domain walls that can be expected to form during a suitable phase transition. In this section we focus on a somewhat different aspect of the problem: what are the statistical properties of the domain walls formed at a phase transition? Are the domain walls formed as little closed spherical structures? Or are they infinite and planar? First we discuss the simple case of a network of Z_2 walls and then the more complicated case of $SU(5) \times Z_2$ walls.

6.6.1 Z_2 network

The properties of the network of Z_2 domain walls at formation have been determined by numerical simulations implementing the "Kibble mechanism." The vacuum in any correlated region of space is determined at random. Then, if there are only two degenerate vacua (call them black and white), there are spatial regions that are in the black phase with 50% probability and others in the white phase. The boundaries between these regions of different phases are the locations of the domain walls (see Fig. 6.6).

Numerical simulations of the Kibble mechanism on a cubic lattice gave the statistics shown in Table 6.1 [74, 159]. The data show that there is essentially one giant connected black cluster. By symmetry there is one connected white cluster. In the infinite volume limit, these clusters are also infinite and their surface areas are infinite. Therefore the topological domain wall formed at the phase transition is infinite.

Figure 6.6   The distribution of two phases (black and white) on a square lattice in two spatial dimensions. Domain walls lie at the interface of the black and white regions.

6.6.2 *SU*(5) *network*

What does the Kibble mechanism predict for $SU(5) \times Z_2$ domain walls? Just as in the Z_2 case, we have to throw down values of the Higgs field on a lattice, assuming that every point on the vacuum manifold is equally likely, and then examine the walls that would form at the interface. In Section 2.2 we have found that there are three kinds of wall solutions in this model and we have labeled the walls by the index q, which can take values 0, 1, or 2. Each kind of wall has the same topology but they have different masses. Each wall type is formed with some probability. Based on the Kibble mechanism, the probability that a certain wall forms is directly related to the number of boundary values that result in the formation of that kind of wall. So we need to evaluate all the boundary conditions that lead to domain walls with a certain value of q.

The space of boundary conditions leading to a given type of domain wall is discussed in Section 2.4. However, similar considerations occur in simpler models and it is helpful to think of the problem in a discrete case, for example the $S_5 \times Z_2$ kinks described in Section 2.5. Take a fixed (discrete) vacuum in one domain. The neighboring domain can be in any other vacuum state with equal probability. There are ten possible states for the neighboring domain. Only one of these gives the $q = 0$ wall, six give the $q = 1$ wall, and three give the $q = 2$ wall. Then the Kibble mechanism implies that the network contains $q = 0, 1, 2$ walls and their number densities are in the ratio $1 : 6 : 3$. This means that the network is dominantly composed of the $q = 1$ wall. However, since the $q = 2$ wall is the lightest wall, the $q = 1$ walls formed by the Kibble mechanism during the phase transition subsequently decay into $q = 2$ walls. We will show some evidence for this two-stage process in Section 6.7.

Similarly we can identify the space of boundary conditions that lead to a particular kind of kink in the $SU(5) \times Z_2$ model. We considered this problem in Section 2.4 and listed the spaces in Table 2.1. From the table we read off that the space of boundary conditions leading to the $q = 0$ kink is zero dimensional, for the $q = 1$ kink it is six dimensional, and is four dimensional for the $q = 2$ kink. Since a six-dimensional space is infinitely bigger than a 0- or a four-dimensional space, the probability of a kink being of the $q = 0$ or $q = 2$ variety is zero, and the probability of the kink being of the $q = 1$ variety is 1.

A subtlety that has not been discussed above is that there is also the possibility that if we lay down Higgs fields randomly, we may get $[\Phi_-, \Phi_+] \neq 0$ (see theorem in Section 2.2). In this case, as described in Section 2.2, there is no static solution to the equations. Then the field configuration evolves toward a static configuration. Our discussion above assumes that such a configuration has been reached, and neighboring domains always have values of Φ that commute. This is not completely satisfactory since there are time scales that are associated with the relaxation and these must be compared to other time scales characterizing the phase transition. This is why a numerical study, such as that in Section 6.7, is needed.

To summarize, the Kibble mechanism predicts that only $q = 1$ domain walls are formed at the $SU(5) \times Z_2$ phase transition. However, we know that the stable variety of walls have $q = 2$, and hence the $q = 1$ walls decay into them. The formation of walls and the conversion of $q = 1$ walls into $q = 2$ walls during a phase transformation in the $SU(5) \times Z_2$ model has not been studied. However, these questions have been addressed in the related $S_5 \times Z_2$ model as we now discuss.

6.7 Formation of $S_5 \times Z_2$ domain wall network

As discussed in the last section, the $q = 1$ domain wall of the $S_5 \times Z_2$ model occupies the largest volume in the space of boundary conditions but the $q = 2$ wall has least energy. Hence there is a tension between "entropy" (number of states) and "energy" (mass of wall). In a phase transition, based on the Kibble argument, we might expect the entropy to be more important. However, the higher energy walls $q = 1$ cannot survive indefinitely and eventually decay into the $q = 2$ walls. One way to study these processes is by direct simulation of the fields as a function of temperature [123, 6, 7].

The simulations are based on a Langevin equation where thermal effects are treated as a noise term in the classical equations of motion together with a damping term. For the $S_5 \times Z_2$ model (Eq. (2.30)) with its four scalar fields, the equations are

$$\left(\partial_t^2 - \nabla^2\right) f_i + V_i + \Delta \, \partial_t f_i = \Gamma_i \tag{6.61}$$

Formation of kinks

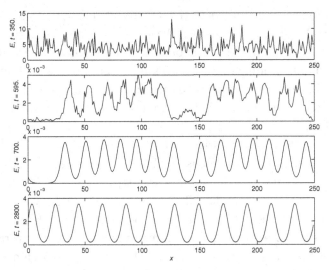

Figure 6.7   Energy density distribution in space at an early time at high temperature (top panel) and then at various times after the phase transition in the lower panels. The last panel shows that the system has relaxed into a stable lattice of kinks.

where $i = 1, \ldots, 4$ and V_i denotes the derivative of V with respect to f_i. If $\Delta = 0 = \Gamma_i$, these equations are simply the classical equations of motion for the f_i. In a thermal system, we imagine that the fields are in contact with a heat bath at temperature T with which energy can be exchanged. Then there can be dissipation which is represented by the Δ term and thermal noise which is represented by the Γ_i term. The dissipation constant Δ is taken to be independent of the temperature but the Γ_i are stochastic and taken to be Gaussian distributed with the following correlation functions

$$\langle \Gamma_i(\mathbf{x}, t) \rangle = 0,$$
$$\langle \Gamma_i(\mathbf{x}, t) \Gamma_j(\mathbf{y}, t') \rangle = 2\Delta \, T \, \delta_{ij} \delta(\mathbf{x} - \mathbf{y}) \delta(t - t') \tag{6.62}$$

The procedure is to solve Eq. (6.61) with any initial condition. The noise and dissipation eventually drive the system to a thermal distribution at temperature T. To mimic the phase transition, the noise is then set to zero. All of a sudden the system has to find a new equilibrium state. This equilibrium state has domain walls and these are located and tracked in the subsequent evolution.

In one spatial dimension, the results are shown in Fig. 6.7. At high temperature the energy distribution is very noisy. After the phase transition, the presence of kinks is clear. During the evolution, some of these kinks annihilate. In the end we are left with a kink lattice.

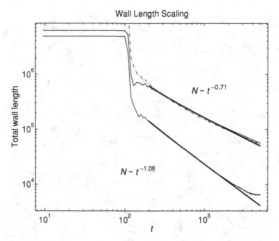

Figure 6.8 Length in walls (denoted by N) in two spatial dimensions against time for the $q = 2$ walls in the $S_5 \times Z_2$ model (upper solid curve) and the single field Z_2 case (lower solid curve). The dashed curve corresponds to the total $S_5 \times Z_2$ wall length measured by counting zeros of the diagonal elements of $\Phi(x)$ and hence includes walls with any value of q. The difference between the solid and dashed curves shows that the initial network consists of a large fraction of $q \neq 2$ walls but then later all the walls decay into the $q = 2$ walls. Comparison with the Z_2 case shows that the $S_5 \times Z_2$ network decays more slowly. (The upturn at the very end in the Z_2 case is due to the finite simulation box.)

Similar numerical simulations have also been done in two spatial dimensions. The total energy in all kinds of walls is plotted as a function of time in Fig. 6.8. The figure also shows the energy in only the $q = 2$ walls as a function of time. The difference of these curves shows that not all walls are of the $q = 2$ variety at formation. Other kinds of walls are present immediately after the phase transition but they must then decay into the least massive $q = 2$ wall.

As discussed in Section 2.8, the $S_5 \times Z_2$ kinks can have nodes in two spatial dimensions (see Fig. 2.6). So we expect a network of domain walls to form after a phase transition in which six or more domain walls are joined at junctions. This is exactly what is seen in simulations (Fig. 6.9). Another feature that is apparent on a closer look at the network is that there are many pairs of walls that are very close to each other. These pairs occur because the unstable $q = 1$ walls eventually decay into two $q = 2$ walls. The forces separating the $q = 2$ walls are exponentially small and so they stay close-by during further evolution.

We have seen that the final state of the $S_5 \times Z_2$ phase transition in one spatial dimension is a lattice of domain walls (Fig. 6.7). In one dimension, it can be argued that a lattice forms with unit probability provided the size of the simulation box is much larger than the wall thickness. In two dimensions, if the spatial extent in one

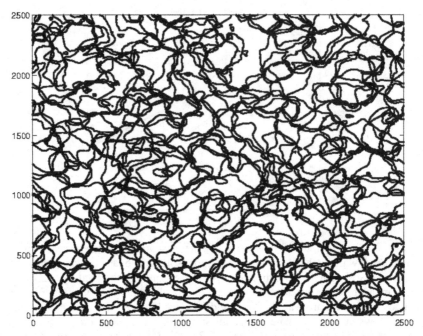

Figure 6.9   Network of $S_5 \times Z_2$ walls in two spatial dimensions soon after the phase transition. The picture looks very similar to the network of (one-dimensional) walls connected to a network of (point-like) strings studied in [130].

direction is smaller than that in the other direction, so that the simulation box is rectangular with periodic boundary conditions, the evolution is very much like in one dimension and a lattice forms once again (see Fig. 6.10). Even on a square two-dimensional simulation box, a domain wall lattice is seen to form with a probability ~ 0.05 [7].

6.8 Biased phase transitions

The existence of domain walls relies only on the existence of discrete vacua. Then it is possible to imagine situations where the degeneracy of the discrete vacua is slightly broken (see Fig. 6.11).[4] Now the probability that the higher energy vacuum is selected during the phase transition in some region is less than $1/2$ and the probability that the lower energy vacuum is selected is larger than $1/2$. This process can again be simulated on a square lattice by throwing down black squares with probability $p < 1/2$. If p is very small, there are only a few black squares and these are disconnected from each other. So the domain walls are small

---

[4] Or perhaps the vacua are exactly degenerate but the likelihood of being in one particular vacuum is slightly larger because of the way in which the system was prepared.

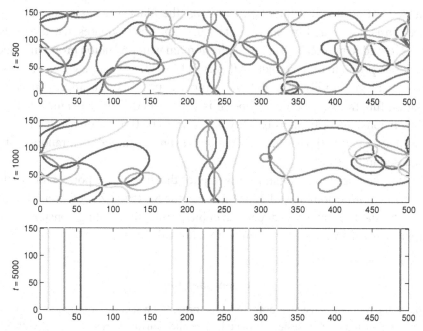

Figure 6.10   Three stages for the domain wall network evolution in a toroidal domain, with dimensions $L_x = 500$ and $L_y = 150$. The different shades correspond
to the five possible charges of the domain walls (see Section 2.7). Note that in the
bottom figure there is a pair of neighboring wall and antiwall of the same type (the
walls just before and after the 300 mark). These later annihilate, leading to a final
stable lattice consisting of ten walls.

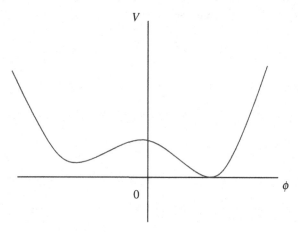

Figure 6.11   An asymmetric well in which the degeneracy of the vacua is slightly
broken.

and closed. At a critical value of p, call it p_c, the black squares connect and the distribution is dominated by one infinite cluster of black squares. Then the black squares are said to percolate. Therefore the domain wall formation problem reduces to the classic problem of "percolation theory" [143, 96, 36] where we are interested in the critical probability and also the critical exponents that appear in various correlation functions as the critical point is approached. On a triangular lattice in two dimensions, the critical probability is known to be 0.5 and on a cubic lattice in three dimensions it is 0.31. The problem may even be studied on a random lattice as discussed in [95].

The analysis for biased domain walls implies that even if the potential is slightly asymmetric, infinite domain walls can form. For the $SU(5) \times Z_2$ potential described in Eq. (2.5), the asymmetry is due to the cubic term with coupling constant, γ. For small but non-zero values of γ, infinite domain walls form.

6.9 Open questions

1. What is the number density of domain walls formed in a second-order phase transition? The question may need to be sharpened since the density keeps changing with time. Also, while the number density is important in cosmology, condensed matter physicists are mainly concerned with scaling laws (critical exponents) since these are expected to be universal. So a sharper question would be in terms of a critical exponent related to the number density (e.g. Eq. (6.60)).
2. Is there a condensed matter system which gives a domain wall lattice?
3. Is there any role for domain wall lattices in (higher dimensional) cosmology?
4. If a domain wall lattice can be generalized to strings and monopoles, do string and monopole lattices form during a phase transition?

7

Dynamics of domain walls

In this chapter we discuss the dynamics of kinks and domain walls first in the zero thickness approximation, and then briefly in the full field theory. The zero thickness approximation can be expected to be valid in the case when all other length scales, such as the radii of curvature of a domain wall, are much larger than the wall thickness.[1] We start by deriving the action for a kink in $1 + 1$ dimensions as this is the simplest case and contains the essential features of the higher dimensional cases. Then we derive the action for a domain wall in $3 + 1$ dimensions and some consequences. In this chapter we ignore gravitational effects which can be quite important in certain situations (see Chapter 8).

7.1 Kinks in $1 + 1$ dimensions

In $1 + 1$ dimensions, if we ignore the structure of the kink, then we expect the kink to behave simply as a massive point particle. Its dynamics are then given by the usual action for a massive point particle

$$S_{1+1} = -M \int d\tau \qquad (7.1)$$

where M is the mass of the kink and $d\tau$ is the line element which may also be written as

$$d\tau = dt \left(g_{\mu\nu} \frac{dX^\mu}{dt} \frac{dX^\nu}{dt} \right)^{1/2} \qquad (7.2)$$

where $g_{\mu\nu}$ is the metric of the space-time background and $X^\mu(t)$ is the location of the kink at time t.

While the action in Eq. (7.1) seems reasonable on physical grounds, there should be a systematic way to derive it starting from the original field theory action of which

---

[1] This expectation is not completely correct since the wall velocity is also important [183, 73].

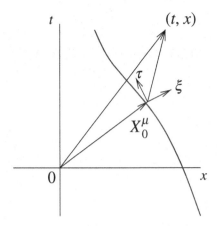

Figure 7.1   The world-line of the kink is represented by the curve. The kink frame coordinates $y^a = (\tau, \xi)$ are defined in the instantaneous rest frame of the kink and are functions of the background coordinates $x^\mu = (t, x)$.

the kink is a solution. Such a derivation should lead to Eq. (7.1) plus corrections that depend on the internal structure of the kink.

To derive the effective action (Eq. (7.1)), the key assumption is that the field profile of the kink is well-approximated by that of the known static kink solution in the instantaneous rest frame of the kink. To proceed with the derivation, we work in "kink frame coordinates" which are denoted by $y^a = (\tau, \xi)$, $a = 0, 1$, as illustrated in Fig. 7.1 (τ is also called the kink world-line coordinate). These coordinates are functions of the background coordinates that are denoted by $x^\mu = (t, x)$, $\mu = 0, 1$.

The kink world-line is given by the position 2-vector $X^\mu = (t, X(t))$. Therefore the vector tangent to the world-line is $T^\mu = N_T(1, \partial_t X)$ where N_T is a normalization factor chosen to enforce

$$g_{\mu\nu} T^\mu T^\nu = 1 \tag{7.3}$$

The unit vector, $N^\mu(\tau)$, orthogonal to the world-line is found by solving

$$g_{\mu\nu} T^\mu N^\nu = 0 \tag{7.4}$$

together with the normalization condition

$$g_{\mu\nu} N^\mu N^\nu = -1 \tag{7.5}$$

In the special case of a Minkowski background, $g_{\mu\nu} = \eta_{\mu\nu} = \mathrm{diag}(1, -1)$, we find $T^\mu = \gamma(1, V)$ where $V \equiv \partial_t X$, and $N^\mu = \gamma(V, 1)$ where $\gamma = 1/\sqrt{1 - V^2}$.

The coordinate τ is along T^μ and ξ is along N^μ. Therefore, in the neighborhood of some fixed point on the world-line, any space-time point can be written as

$$x^\mu = X_0^\mu + \tau T_0^\mu + \xi N_0^\mu \equiv X^\mu(\tau) + \xi N^\mu(\tau_0) \tag{7.6}$$

where the subscript 0 refers to the fixed point on the world-line. Since the energy density in the fields vanishes far from the kink, only the neighborhood of the world-line is relevant for deriving the effective action. Hence ξ is small and to lowest order we can replace τ_0 in the last term by τ to get

$$x^\mu = X^\mu(\tau) + \xi N^\mu(\tau) \tag{7.7}$$

With the coordinate transformation in Eq. (7.7), the world-line metric can be written in the y^a coordinate system

$$h_{ab} = g_{\mu\nu} \partial_a x^\mu \partial_b x^\nu \tag{7.8}$$

Therefore

$$
\begin{aligned}
h_{00} &= g_{\mu\nu}(\partial_\tau X^\mu + \xi \partial_\tau N^\mu)(\partial_\tau X^\nu + \xi \partial_\tau N^\nu) \\
&= g_{\mu\nu} \partial_\tau X^\mu \partial_\tau X^\nu + O(\xi) \\
h_{01} &= g_{\mu\nu}(\partial_\tau X^\mu + \xi \partial_\tau N^\mu) N^\nu = O(\xi) \\
h_{11} &= g_{\mu\nu} N^\mu N^\nu = -1
\end{aligned}
$$

where we have used the orthogonality of $\partial_\tau X^\mu \propto T^\mu$ and N^μ, and the normalization of N^μ. So the determinant of h_{ab} is

$$h = -g_{\mu\nu}(X^\mu) \partial_\tau X^\mu \partial_\tau X^\nu + O(\xi) \tag{7.9}$$

where we have also expanded the background metric around the kink location.

Next we write,

$$\phi(x^\mu) = \phi_0(y^a) + \psi(y^a) \tag{7.10}$$

where ϕ_0 is the static kink profile function in the kink frame coordinates. For example, in the case of the Z_2 kink, $\phi_0 = \eta \tanh(\sqrt{\lambda/2}\, \eta\xi)$ (see Eq. (1.9)). The field ψ is the departure of the true field configuration from the static kink profile ϕ_0. The assumption is that the contribution of ψ to the action is small and hence ψ can be used as a parameter for a perturbative expansion.

Now the field theory action is

$$S = \int d^2x \sqrt{-g} L[\phi, \dot\phi; g_{\mu\nu}] \tag{7.11}$$

in terms of the Lagrangian density L and $g = \mathrm{Det}(g_{\mu\nu})$. The metric is taken to be a fixed background and the gravitational effects of the wall are ignored. The full problem of gravitating domain walls is significantly more complicated at a technical level [21].

Now we write this action in the kink frame coordinates to get

$$S = S_0 + O(\xi, \psi) \tag{7.12}$$

with

$$S_0 = \int d\tau d\xi \sqrt{|h|} \, L[\phi_0(\xi), \dot{\phi}_0(\xi); h_{ab}]$$

$$= \int d\tau \sqrt{|h|} \int d\xi \, L[\phi_0(\xi), 0; h_{ab}]$$

$$= -M \int d\tau \sqrt{|h|} \tag{7.13}$$

where M is the mass of the kink. The last equality follows since the solution is static and hence the Lagrangian density equals the energy density up to a sign. The integration of the Lagrangian density then gives the $-M$ factor. The effective action is therefore the action for a point particle, simply given by the length of the world-line. This result can easily be extended to walls (and strings) propagating in higher dimensions, and the leading term in the effective action is proportional to the world volume. Such geometric effective actions are often referred to as "Nambu-Goto actions." Even if the self-gravity of the domain wall is taken into account, the dominant contribution to the effective action is still the Nambu-Goto action [21].

The next-to-leading order terms in the effective action, denoted by $O(\xi, \psi)$ in Eq. (7.12), have been discussed for domain walls in [138, 21, 73, 28], building on the earlier analysis for strings [57, 105, 72]. The first-order corrections in both ψ and ξ vanish because the field ϕ_0 is a solution of the equation of motion and hence the action is an extremum at ϕ_0. The lowest non-trivial corrections come at second order in ξ and ψ. An alternative approach to studying domain wall dynamics has been developed in [9].

Finally we remark that the parameter τ can be chosen arbitrarily. Any other world-line coordinate, $\tau'(\tau)$, leaves the effective action invariant. This fact is called "reparametrization invariance" of the action.

7.2 Walls in 3 + 1 dimensions

The location of a domain wall, $X^\mu(\tau, \zeta, \chi)$, is described by three world-volume coordinates $y^a = (\tau, \zeta, \chi)$. Any point, x^μ, can now be written in terms of the "wall frame coordinates" (see Eq. (7.6) and Fig. 7.2)

$$x^\mu = X^\mu(\tau, \zeta, \chi) + \xi N^\mu(\tau, \zeta, \chi) \tag{7.14}$$

where N^μ is the normal to the wall.

The derivation of the Nambu-Goto action proceeds exactly as in the kink case of the last section and we get

$$S_0 = -\sigma \int d^3\rho \sqrt{|h|} \tag{7.15}$$

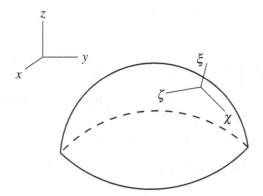

Figure 7.2 A curved section of a domain wall is shown. The world-sheet coordinates are (τ, ζ, χ, ξ) while those in the ambient space ("bulk") are (t, x, y, z).

where σ is the energy per unit area (tension) of the wall, the integral is over the wall world volume, $h = \mathrm{Det}(h_{ab})$, and the world-volume metric is

$$h_{ab} = g_{\mu\nu}(X^\rho)\partial_a X^\mu \partial_b X^\nu \tag{7.16}$$

where $a, b = \tau, \zeta, \chi$. Note that the determinant of h_{ab} is positive for the kink in $1 + 1$ dimensions and also the domain wall in $3 + 1$ dimensions.

The major difference between the kink in $1 + 1$ dimensions and the domain wall is that the wall can be curved, and so the profile ϕ_0, which only applies to planar walls, does not solve the equation of motion. For example, as the wall moves, it accelerates and emits radiation. The radiation part must be treated as a perturbation. However, the analysis is conceptually the same as for the kink and the derivation may be found in [138, 21, 73, 28].

From the Nambu-Goto action for the domain wall, we can derive the equations of motion. The variation of S_0 involves the variation of $h = \mathrm{Det}(h_{\alpha\beta})$. This follows from the identity (see Appendix E)

$$\delta \ln \mathrm{Det} \mathbf{M} = \mathrm{Tr}(\mathbf{M}^{-1}\delta \mathbf{M}) \tag{7.17}$$

valid for any invertible matrix $\mathbf{M}$. Applying this identity to the matrix h_{ab} we get

$$\delta h = h h^{ab} \delta h_{ab} \tag{7.18}$$

where h^{ab} is the inverse of h_{ab} so that

$$h^{ab} h_{bc} = \delta^a_c \tag{7.19}$$

Therefore the variation of S_0 is

$$\delta S_0 = -\frac{\sigma}{2} \int d^3\rho \sqrt{|h|} h^{ab} \delta h_{ab} \tag{7.20}$$

We obtain the wall equation of motion by requiring $\delta S_0 = 0$ together with the definition of h_{ab} in Eq. (7.16)

$$\frac{1}{\sqrt{|h|}}\partial_a(\sqrt{|h|}h^{ab}\partial_b X^\sigma) = \Gamma^\sigma_{\mu\nu}h^{ab}\partial_a X^\mu \partial_b X^\nu \tag{7.21}$$

where the Christoffel symbol is defined by the background metric $g_{\mu\nu}$

$$\Gamma^\sigma_{\mu\nu} = \frac{g^{\sigma\rho}}{2}(\partial_\nu g_{\rho\mu} + \partial_\mu g_{\rho\nu} - \partial_\rho g_{\mu\nu}) \tag{7.22}$$

In the special case of a Minkowski background metric, the Christoffel symbol vanishes and

$$\frac{1}{\sqrt{|h|}}\partial_a(\sqrt{|h|}h^{ab}\partial_b X^\sigma) = 0 \tag{7.23}$$

Using Eq. (7.17), the determinant h can be eliminated and the equation of motion can be written as

$$\partial_a(h^{ab}\partial_b X^\sigma) + \frac{1}{2}h^{cd}\partial_a h_{cd}\, h^{ab}\partial_b X^\sigma = 0 \tag{7.24}$$

The equation of motion for a wall is highly non-linear because h_{ab} itself is defined as a quadratic in derivatives of X^μ. One way to simplify the equations is to choose convenient coordinates. This is possible because the equations of motion of the wall are reparametrization invariant, i.e. we are free to choose any world-volume coordinates (τ, ζ, χ). A similar situation occurs for strings that have a $1+1$ dimensional world sheet. There, by a choice of coordinates, the equation of motion can be converted to a simple wave equation in $1+1$ dimensions together with some quadratic constraints that can be solved quite generally. In the case of the domain wall, however, no such convenient choice of coordinates is known and the equations have not been solved in general. Only a few special solutions are known. Of these, static solutions subject to suitable boundary conditions have minimal surface area, and these have been extensively studied in the mathematics literature e.g. [115].

In a realistic setting, the dynamics of the walls are affected by inter-kink forces, by the interaction of any surrounding particles, the gravitational field of the wall, and the evolution of the background space-time. In addition, there are collisions between different walls, leading to intercommuting (Section 3.8), and annihilation of walls and antiwalls. If there are zero modes on the walls as described in Chapter 5, they could also carry charges and currents and this would introduce other interactions.

7.3 Some solutions

In $1+1$ dimensions the kink moves like a point particle of mass M. The dynamics are richer in $3+1$ dimensions where a closed domain wall can oscillate and move in complicated ways. The Nambu-Goto action is valid when the radii of curvature of the wall and the separation of different sections of wall are both large compared to the thickness of the wall. In addition, the velocity of the wall (in the center of mass frame) should be small. (See Section 7.3.3 for the criterion in the case of collapsing spherical domain walls.) When these conditions are not met, the only way to proceed is to consider the dynamics using the underlying field theory. In this section, we ignore field theory effects and describe some solutions to the Nambu-Goto action.

7.3.1 Planar solutions: traveling waves

A planar domain wall in the $z = 0$ plane is given by

$$X^\mu(\tau, \zeta, \chi) = (\tau, \zeta, \chi, 0) \tag{7.25}$$

Next consider a planar domain wall with some ripples

$$X^\mu(\tau, \zeta, \chi) = (\tau, \zeta, \chi, z(\tau, \zeta, \chi)) \tag{7.26}$$

The function z describes the ripples and we would like solutions for z.

For the wall in Eq. (7.26), the world-volume metric is

$$h_{ab} = \eta_{ab} - \partial_a z \partial_b z \tag{7.27}$$

where

$$\eta_{ab} = \mathrm{diag}(1, -1, -1) \tag{7.28}$$

Inverting h_{ab} is not simple, but inverting η_{ab} is. So consider the "trial" inverse metric

$$\tilde{h}^{bc} = \eta^{bc} + \eta^{bd}\partial_d z\, \eta^{ce}\partial_e z \tag{7.29}$$

Then by evaluating $h_{ab}\tilde{h}^{bc}$, it can be seen that $\tilde{h}^{bc}$ is the correct inverse metric provided

$$\eta^{ab}\partial_a z \partial_b z = 0 \tag{7.30}$$

Now we can use Eq. (7.24) and the constraint (7.30) to get the equation of motion for the function $z(\tau, \zeta, \chi)$

$$\partial^a \partial_a z = 0 \tag{7.31}$$

Hence any function that satisfies Eqs. (7.31) and (7.30) extremizes the Nambu-Goto action for a domain wall.

Solutions of Eqs. (7.31) and (7.30) have been discussed in [58]. The constraint condition implies that the solution must necessarily be time-dependent. A class of solutions is obtained by noting, for example, that $z = f(\tau - \zeta)$ solves the equation of motion and also the constraint for any choice of function f. This corresponds to a pulse of arbitrary shape on a planar domain wall that propagates in the $+x$

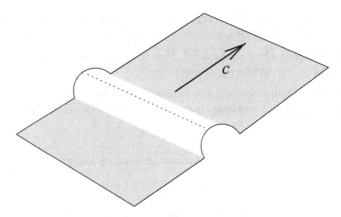

Figure 7.3   Sketch of a traveling wave on a planar domain wall. The pulse propagates at the speed of light along the wall.

direction at the speed of light. Similarly

$$z = f(\tau \pm (n_1 \zeta + n_2 \chi)), \qquad n_1^2 + n_2^2 = 1 \qquad (7.32)$$

is a solution for any unit vector (n_1, n_2). These solutions are known as "traveling waves" (see Fig. 7.3).

Other solutions of the wave equation (Eq. (7.31)) are also known – for example, circular waves – but these do not satisfy the constraint equation and/or have singularities.

7.3.2 Axially symmetric walls

Here we look for a static wall solution in a Minkowski background. The (Cartesian) coordinates of the wall take the form

$$X^\mu(\tau, \theta, \lambda) = (\tau, R(\lambda)\cos\theta, R(\lambda)\sin\theta, \lambda) \qquad (7.33)$$

with $\eta_{\mu\nu} = \mathrm{diag}(1, -1, -1, -1)$. The wall metric is seen to be

$$h_{ab} = \mathrm{diag}(1, -R^2, -(1 + R'^2)) \qquad (7.34)$$

where R' is the derivative of R with respect to λ. The equation of motion, Eq. (7.21), then leads to

$$\frac{\mathrm{d}}{\mathrm{d}\lambda}\left(\frac{R}{\sqrt{1 + R'^2}}\right) = 0, \qquad \frac{\mathrm{d}}{\mathrm{d}\lambda}\left(\frac{RR'}{\sqrt{1 + R'^2}}\right) = \sqrt{1 + R'^2} \qquad (7.35)$$

with the solution

$$R(\lambda) = \frac{1}{\alpha}\cosh(\alpha\lambda) \qquad (7.36)$$

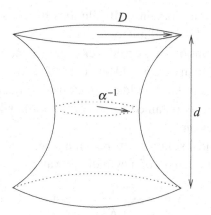

Figure 7.4   Sketch of a catenoid solution.

where α is a parameter, $\tau = t$, $\lambda = z$ and θ is the angle in cylindrical coordinates. Equation (7.36) describes a one-parameter family of static, axially symmetric, domain wall solutions (see Fig. 7.4).

The solution in Eq. (7.36) is a catenoid that is seen in soap films which, like domain walls, also minimize their surface area [23]. Experiments with soap films are done with two parallel circular rings, each of diameter D, placed a certain distance, d, apart. Then the soap film forms a catenoid for $d/D < 0.66$ [117]. Actually there are two catenoid solutions for $d/D < 0.66$ since the relation $\alpha D = \cosh(\alpha d/2)$ has two solutions for α for fixed values of D and d in this regime. A third solution, which consists of two disconnected disks circumscribed by each of the circular rings also exists. For larger values of the separation-to-diameter ratio, d/D, the two-disk solution has less surface area than the catenoid solutions, and the catenoid can pinch off and minimize its area by transforming to the two disks. It seems reasonable to assume that the soap film analysis also applies to the domain wall.

The catenoid is a static solution of the Nambu-Goto equations of motion. It could happen that the catenoid is not a solution of the field equations. A simple example of a solution to the Nambu-Goto equations that does not solve the field equations can be constructed quite easily. Two parallel planar walls (a wall and an antiwall) form a solution to the Nambu-Goto equations but, since these walls have an exponentially small attractive force, they do not form a solution to the field equations. However, by fixing the boundary conditions (as in the soap film case by the rings), the catenoid solution for domain walls has been constructed numerically by solving the equations of motion for the scalar field in the Z_2 model (Sutcliffe, P., 2005, private communication). The stability of the catenoid solution to the Nambu-Goto equations is an open question (Section 7.7).

Quite complicated static domain wall solutions have also been studied in the context of quasicrystals [137] and microemulsions [70].

In addition to static solutions, we could seek time-dependent solutions with axial symmetry. The simplest such case would be a cylindrical domain wall whose radius is a function of time. The radius would contract, pass through zero, and then grow again. A similar solution is obtained for spherical walls which we discuss more explicitly in the next section.

To obtain the cylindrical solution, we note that energy is conserved during collapse. The energy per unit length of a cylindrical wall is

$$\Lambda = \frac{\sigma \, 2\pi \, R}{\sqrt{1 - \dot{R}^2}} = \text{constant} \tag{7.37}$$

where σ is the energy per unit area of the wall, R is the radius of the cylinder at time t, and an overdot denotes differentiation with respect to t. The square root factor in the denominator takes care of the Lorentz boost.

The conservation of energy (i.e. constancy of Λ), immediately leads to the solution

$$R(t) = R_0 \cos\left(\frac{t}{R_0}\right) \tag{7.38}$$

where $R_0 = \Lambda/\sigma 2\pi$ is the radius when the wall is at rest.

7.3.3 Spherical walls

Our final example of domain wall solutions is with a spherical ansatz

$$X^\mu(\tau, \theta, \phi) = (\tau, R(\tau)\hat{\mathbf{r}}) \tag{7.39}$$

where

$$\tau = t, \quad \hat{\mathbf{r}} = (\sin\theta \cos\phi, \sin\theta \sin\phi, \cos\theta) \tag{7.40}$$

and θ, ϕ are spherical angular coordinates. The space-time metric is $\eta_{\mu\nu} = \text{diag}(1, -1, -1, -1)$.

We now find

$$h_{ab} = \text{diag}(1 - \dot{R}^2, -R^2, -R^2 \sin^2\theta) \tag{7.41}$$

where overdots denote derivatives with respect to τ. After some algebra, from Eq. (7.21) we obtain the equation of motion

$$\ddot{R} = -\frac{2}{R}(1 - \dot{R}^2) \tag{7.42}$$

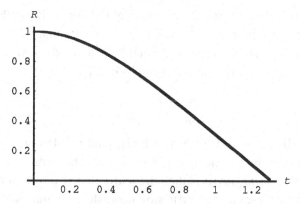

Figure 7.5 Radius of a collapsing spherical domain wall against time in the thin wall approximation. The coordinates in the plot are in units of the maximum radius of the wall.

For $R \neq 0$, $\dot{R}^2 \neq 0$, 1 this can also be written as

$$\frac{d}{d\tau}\left(\frac{R^2}{\sqrt{1 - \dot{R}^2}}\right) = 0 \tag{7.43}$$

which implies

$$4\pi\sigma \frac{R^2}{\sqrt{1 - \dot{R}^2}} = M \tag{7.44}$$

where M is a constant of motion, to be identified with the mass of the spherical domain wall (σ is the mass per unit area of the wall).

The solution can be written in terms of the elliptic integral of the first kind

$$\int_{x_*}^{x} \frac{dx}{\sqrt{1 - x^4}} = \pm\frac{\tau - \tau_0}{R_0} \tag{7.45}$$

where

$$R_0^2 \equiv \frac{M}{4\pi\sigma}, \qquad x \equiv \frac{R}{R_0} \tag{7.46}$$

R_0 has the interpretation of being the radius when the wall is at rest and x_* is the value of x at some initial time τ_0. The sign in Eq. (7.45) is chosen according to whether one is interested in the expanding or contracting solution. The radius of a collapsing spherical domain wall is plotted in Fig. 7.5.

The behavior of perturbations on the spherical domain wall has been studied in [182]. The result is that at late times the ratio of the perturbation amplitude divided by the radius of the spherical wall, grows as $1/R$ as the wall collapses.

In the Nambu-Goto description, the spherical domain wall oscillates about the center. However, the solution is only valid as long as the thin wall approximation holds. By comparing various terms in the field equations of motion, the thin wall approximation is seen to break down when [183, 73]

$$\frac{R}{R_0} \sim \left(\frac{w}{R_0}\right)^{1/3} \tag{7.47}$$

where w is the wall thickness. This relation is also confirmed by numerically solving the equation of motion in the field theory [183]. In [73], the leading order corrections owing to the thickness and gravity of the spherical domain wall are included, with the conclusion that both these effects tend to slow down the dynamics. The Nambu-Goto action also becomes inadequate owing to radiative losses. As the wall collapses, we expect energy losses owing to radiation and eventually annihilation of the domain wall into radiation. We discuss these processes further in Section 7.5.

The collapse of a zero thickness spherical domain wall is prevented if the background space-time is expanding. Static solutions are obtained if the background is expanding at a constant rate, as in de Sitter space. In a particular coordinate system, the line element for de Sitter space becomes time independent

$$ds^2 = f(r)dt^2 - f^{-1}(r)dr^2 - r^2(d\theta^2 + \sin^2\theta d\phi^2) \tag{7.48}$$

where $f(r) = 1 - H^2 r^2$ and H is a constant corresponding to the expansion rate. Following the analysis of [16] for a circular string, the action for a spherical domain wall in the zero thickness limit is

$$S = -4\pi\sigma \int dt\, R^2 \sqrt{f - \frac{\dot{R}^2}{f}} \tag{7.49}$$

where $R(t)$ is the radius of the spherical wall and $f = f(R)$. Extremization of this action leads to the first integral

$$\dot{R}^2 - f^2 + \epsilon^{-2} R^4 f^3 \equiv \dot{R}^2 + V(R) = 0 \tag{7.50}$$

where $\epsilon = E/4\pi\sigma$ and E is a constant (the first integral). For a static solution we need both $V(R) = 0$ and $V'(R) = 0$ where prime denotes derivative with respect to R. These conditions give the static solution

$$R = H^{-1}\sqrt{\frac{2}{3}} \tag{7.51}$$

with

$$E = \frac{4\pi\sigma}{H^2} \frac{2}{3\sqrt{3}} \tag{7.52}$$

The potential $V(R)$ is a maximum at the location of this solution and therefore the solution is unstable. The instability can be understood without calculation. If the radius of the wall is perturbed to be a little smaller than the value at the solution, the effects of Hubble expansion are weaker while the force owing to curvature is stronger, and so the wall collapses. On the other hand, if the radius is perturbed to be a little larger than the solution value, the expansion effect is stronger while the curvature force is weaker, and the wall expands to yet greater radii.

Planar and spherical domain walls in de Sitter space have been considered in the full field theory in [17, 18]. It is found [17] that instanton solutions describing the nucleation of spherical domain walls exist only when the thickness of the wall is less than $H^{-1}/\sqrt{2}$. This result is also relevant to the problem of finding static thick spherical domain walls in de Sitter space, since an instanton solution can exist only if the static domain wall solution exists (though the converse may not hold). Hence spherical domain wall solutions of the field theory in de Sitter space exist if the domain wall thickness is less than $H^{-1}/\sqrt{2}$.

7.4 Solutions in field theory: traveling waves

The traveling wave solutions discussed in Section 7.3.1 in the zero thickness approximation are also exact solutions to the field equations of motion [160, 161].

Consider the field

$$\phi(t, \mathbf{x}) = \phi_0(z - z_0(t, x, y)) \tag{7.53}$$

where $\phi_0(z)$ is the classical solution for a domain wall in the $z = 0$ plane. We now insert this ansatz in the field theory equation of motion. A little algebra shows that the ansatz is a solution provided

$$\partial_a \partial^a z_0 = 0, \quad (\partial_a z_0)^2 = 0 \tag{7.54}$$

where $a = t, x, y$. These are the same equations obtained above for planar solutions to the Nambu-Goto equations (Eqs. (7.31) and (7.30)). As discussed there, the only non-singular solutions to these equations have the form of traveling waves e.g.

$$z_0(t, x, y) = f(t \pm x, y) \tag{7.55}$$

Hence traveling waves are solutions to the field equations and do not dissipate owing to radiation.[2]

---

[2] It can be shown that traveling waves do not dissipate even when they are considered in quantum field theory [46].

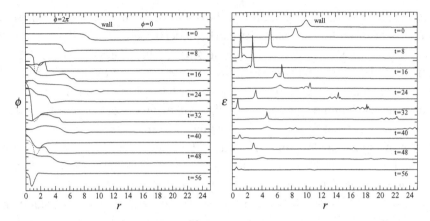

Figure 7.6   Collapse of a spherical sine-Gordon domain wall. The curves in the left-hand plots show the field as a function of radial distance for several different times. The right-hand plot shows the corresponding energy density distributions. [Figure reprinted from [183].]

7.5 Spherical domain walls: field theory

We have seen in Section 3.7 that the collision of a kink and an antikink in $1 + 1$ dimensions leads to chaotic dynamics. The kinks bounce back for certain velocities while for other velocities, both smaller and larger, they annihilate. So we might expect the dynamics of a collapsing spherical domain wall to show similar features. Numerical simulations of the sine-Gordon model show that a collapsing spherical domain wall does not radiate very much energy until it becomes very small (of order the thickness of the wall), then emits a large amount of radiation, then bounces back to form an expanding spherical domain wall (though with less energy than the initial configuration), which then reverses and collapses again (see Fig. 7.6). Simulations of a $\lambda\phi^4$ spherical domain wall, however, do not show any bounce back [183].

7.6 Kink lattice dynamics (Toda lattice)

In Section 6.6.2 we have seen that a phase transition can lead to the formation of a lattice of kinks (Fig. 6.7). What happens if one of the kinks in a lattice collides with a neighboring kink? The interaction potential between neighboring kinks decays exponentially with distance and energy conservation implies that the collision is perfectly elastic. The momentum of the incoming kink is transferred to the target kink [123]. These properties are exactly those assumed for a chain of masses in what is called a "Toda lattice" [155]. The many beautiful properties of a Toda lattice apply to the (one-dimensional) lattice of kinks as well. For example, there are soliton solutions that run along the Toda lattice. So there are also solitons in the dynamical modes of the kink lattice i.e. solitons in the dynamics of solitons!

7.7 Open questions

1. Are there closed domain walls in three dimensions that do not self-intersect as they oscillate? What happens in higher dimensions?
2. Can one show analytically that walls must intercommute on intersection?
3. When traveling waves on domain walls collide, they dissipate some of their energy. Find the energy radiated. Find the energy that goes into excitations of the bound state in the case of the Z_2 wall.
4. Analyze the catenoid domain wall solution and its stability.

8

Gravity and cosmology of domain walls

Domain walls resulting from a symmetry breaking in the early universe could have novel and dramatic gravitational and cosmological consequences.

We first derive the gravitational effects of a planar domain wall, describing the different ways to view the system. Then we discuss spherical walls as an example of curved domain walls. To discuss the cosmological consequences, it is necessary to have a picture of domain wall formation in the cosmological context. With the background of Chapter 6 we discuss the formation of the wall network in cosmology, then the evolution and cosmological implications. We end by reviewing the cosmological constraints on domain walls and the few possible ways around the constraints.

8.1 Energy-momentum of domain walls

The energy-momentum tensor for a scalar field with potential $V(\phi)$ is given in Eq. (1.39)

$$T_{\mu\nu} = \partial_\mu \phi \partial_\nu \phi - g_{\mu\nu} \left\{ \frac{1}{2} (\partial_\alpha \phi)^2 - V(\phi) \right\} \tag{8.1}$$

In the thin-wall limit, varying the Nambu-Goto action (Eq. (7.15)) gives the energy-momentum tensor

$$T^{\mu\nu} \Big|_{\mathrm{NG}} = \frac{\sigma}{\sqrt{-g}} \int d^3\rho \sqrt{|h|} \, h^{ab} \partial_a X^\mu \partial_b X^\nu \, \delta^{(4)}(x^\mu - X^\mu) \tag{8.2}$$

where X^μ is the location of the wall. For a planar wall located at $x = 0$ in flat space-time, this gives

$$T^{\mu\nu} \Big|_{\mathrm{NG,plane}} = \sigma(1, 0, -1, -1)\delta(x) \tag{8.3}$$

For a planar wall, including self-gravity, the energy-momentum tensor can be explicitly written once we have chosen a suitable ansatz for the metric (see Eq. (8.4) below).

8.2 Gravity: thin planar domain walls

The gravitational effects of a planar domain wall have been found in the thin-wall limit in [167, 169, 80] using the vacuum solutions found in [154]. The thin-wall limit simplifies the analysis because then there is no need to solve the field equations of motion. All the energy-momentum is localized on the thin domain wall and so only the vacuum Einstein equations need to be solved on either side of the wall. The presence of the wall shows up in matching the vacuum solutions on the two sides of the wall i.e. implementing the "junction conditions." Such a matching is facilitated by using the Gauss-Codazzi formalism [81] and this has been done in [80]. Here we derive the metric of a domain wall without going through the general Gauss-Codazzi formalism, following the derivation in [169] instead.

A planar domain wall located in the $x = 0$ plane has rotational symmetry in this plane. Further we expect space-time symmetry under $x \to -x$. Under these conditions the form of the line element can be taken to be [154]

$$ds^2 = e^{2u}(+dt^2 - dx^2) - e^{2v}(dy^2 + dz^2) \tag{8.4}$$

where u and v are functions of t and $|x|$. Note that the possibility that the metric is time-dependent has been retained.

In the thin-wall limit, there is no energy-momentum off the wall and so the energy-momentum tensor, $T_{\mu\nu}$, vanishes everywhere except on the wall. Therefore only the vacuum Einstein equations, $R_{\mu\nu} = 0$, where $R_{\mu\nu}$ is the Ricci tensor, need be solved. The solution for $x > 0$ is

$$e^{2v} = f(t + x) + g(x - t) \tag{8.5}$$

$$u = -\frac{1}{4} \ln(f + g) + h(x + t) + k(x - t) \tag{8.6}$$

where the functions f, g, h, and k satisfy

$$f'' - g'' - 2f'h' + 2g'k' = 0 \tag{8.7}$$

$$f'' + g'' - 2f'h' - 2g'k' = 0 \tag{8.8}$$

where primes denote derivatives with respect to x. The solution for $x < 0$ can be obtained by symmetry since u and v are functions of $|x|$.

Next we solve the Einstein equations, $T_{\mu\nu} = G_{\mu\nu}/8\pi G$, where $G_{\mu\nu}$ is the Einstein tensor calculated for the metric in Eq. (8.4). This leads to

$$T_0^0 = \frac{1}{4\pi G} v_0' e^{-2u_0} \delta(x)$$
$$T_1^1 = 0$$
$$T_2^2 = T_3^3 = -\frac{1}{8\pi G}(u_0' + v_0') e^{-2u_0} \delta(x) \tag{8.9}$$

where $u_0 = u(t, x = 0+)$, $v_0 = v(t, x = 0+)$.

In general, u_0, u_0', and v_0' are time-dependent, and so these expressions for T_ν^μ are also time-dependent. However, the energy-momentum tensor for the wall should be time-independent. This gives us the constraint that the functions f, g, h, and k must be chosen so that u_0, u_0', and v_0' are time-independent. Then the only possible choice for the functions (for $x > 0$) that also satisfy Eq. (8.8) is

$$f = 0, \qquad g = e^{K(t-x)}$$
$$h = -\frac{K}{4}(t + x), \qquad k = \frac{K}{2}(t - x) \tag{8.10}$$

where

$$K = 4\pi G\sigma \tag{8.11}$$

The corresponding functions for $x < 0$ are

$$f = e^{K(t+x)}, \qquad g = 0$$
$$h = \frac{K}{2}(t + x), \qquad k = -\frac{K}{4}(t - x) \tag{8.12}$$

Then the domain wall line element is

$$ds^2 = e^{-K|x|}[dt^2 - dx^2 - e^{Kt}(dy^2 + dz^2)] \tag{8.13}$$

which can also be put in the commonly encountered form

$$ds^2 = (1 - \kappa|X|)^2 dt^2 - dX^2 - (1 - \kappa|X|)^2 e^{2\kappa t}(dy^2 + dz^2) \tag{8.14}$$

where $\kappa = 2\pi G\sigma$ via the coordinate transformation

$$|X| = \frac{1}{\kappa}(1 - e^{-\kappa|x|}) \tag{8.15}$$

8.3 Gravitational properties of the thin planar wall

On spatial slices of constant X ($X = X_0$) the metric of Eq. (8.14) takes the form

$$ds_3^2 = d\bar{t}^2 - e^{2\bar{\kappa}\bar{t}}(d\bar{y}^2 + d\bar{z}^2) \tag{8.16}$$

where overbars denote that the coordinates have been rescaled by the factor $(1 - \kappa|X_0|)$ and $\bar{\kappa} = \kappa/(1 - \kappa|X_0|)$. The three-dimensional line element of Eq. (8.16) shows that space-like slices of constant X are expanding exponentially fast, just as in an inflationary space-time.

The inflationary nature of the metric can be understood from the viewpoint of an observer living on the wall who is blind to the coordinate normal to the wall. From such an observer's perspective, the space-time is filled with vacuum energy, as given by the energy-momentum tensor of Eq. (8.3), and hence is inflating.

Next we examine the metric on spatial slices obtained by setting $y = y_0$, $z = z_0$

$$ds^2 = (1 - \kappa|X|)^2 dt^2 - dX^2 \qquad (8.17)$$

This is the metric of $1 + 1$ dimensional Rindler space-time, which is Minkowski space-time written in the rest frame coordinates of a uniformly accelerated observer with acceleration $a = 1/\kappa$ *away* from the wall which is located at $X = 0$. To see this, use the coordinate transformation

$$\tau = \frac{(1 - \kappa|X|)}{2\kappa}(e^{\kappa t} - e^{-\kappa t})$$

$$\xi = \frac{(1 - \kappa|X|)}{2\kappa}(e^{\kappa t} + e^{-\kappa t}) \qquad (8.18)$$

In these coordinates the Rindler line element is of Minkowski form

$$ds^2 = d\tau^2 - d\xi^2 \qquad (8.19)$$

Now note that

$$\xi^2 - \tau^2 = \left(\frac{1}{\kappa} - |X|\right)^2 \qquad (8.20)$$

Therefore the world line of a particle at fixed X is a hyperboloid in Minkowski space-time, which describes a particle moving at constant acceleration. In particular, the wall located at $X = 0$ has acceleration $1/\kappa$. Therefore an inertial observer sees the wall accelerating away with acceleration $1/\kappa$. From the perspective of an observer on the wall, all particles are repelled from the wall.

In the Rindler space metric there is a horizon at $|X| = 1/\kappa$. It is clear from the coordinate transformation given above, this is a coordinate singularity since the space-time is equivalent to Minkowski space-time.

As discussed in [80] the full domain wall metric in Eq. (8.14) can also be brought to Minkowski form. This shows explicitly that the domain wall space-time is flat everywhere except on the wall itself. As in the reduced metric of Eq. (8.17), in the

Minkowski coordinates (t_M, x_M, y_M, z_M) the wall is located at (see Eq. (8.20))

$$x_M^2 + y_M^2 + z_M^2 = t_M^2 + \frac{1}{\kappa^2} \tag{8.21}$$

Hence, in the coordinates where the metric is Minkowski, the wall is spherical with time-dependent radius that decreases (for $t_M < 0$) until it gets to $1/\kappa$ at $t_M = 0$ and then bounces back. This behavior does not depend on which side of the wall the observer is located. Both see the wall accelerating away from them with constant acceleration $1/\kappa$.

There is an intuitive way to see that the wall's gravity must be repulsive. In the weak field approximation, the gravitational potential of the wall is proportional to $\rho + p_1 + p_2 + p_3$ where ρ is the energy density of the wall and p_i are the pressure components of the energy-momentum tensor. From the energy-momentum tensor in Eq. (8.3) we have $p_2 = p_3 = -\rho$ and $p_1 = 0$. Therefore $\rho + p_1 + p_2 + p_3 = -\rho < 0$ instead of the positive value obtained for matter without pressure. Therefore the gravitational potential is repulsive instead of being attractive.

Since the metric is Minkowski in the (t_M, x_M, y_M, z_M) coordinates, geodesics are given by

$$x_M^\mu(t_M) = x_0^\mu + u^\mu (t_M - t_0) \tag{8.22}$$

where, x_0^μ is the position of the particle at time $t_M = t_0$, and u^μ is the (constant) velocity vector.

8.4 Gravity: thick planar wall

Here we consider the gravitational field of a *thick* domain wall i.e. taking both the scalar field and Einstein equations into account.

The Einstein equations are

$$G_{\mu\nu} = 8\pi G T_{\mu\nu}$$
$$= 8\pi G \left[\partial_\mu \phi \partial_\nu \phi - g_{\mu\nu} \left\{ \frac{1}{2} (\partial_\alpha \phi)^2 - V(\phi) \right\} \right] \tag{8.23}$$

where we have used $T_{\mu\nu}$ from Eq. (8.1). The scalar field equation is

$$\nabla_\mu \nabla^\mu \phi + V'(\phi) = 0 \tag{8.24}$$

where ∇_μ is the covariant derivative.

These equations have been solved in [181] for the case when $16\pi G \eta^2 \ll 1$ where η is the vacuum expectation value of the field ϕ. The line element outside the thick wall is still given by Eq. (8.14) and there are no qualitative new effects.

The case when $16\pi G\eta^2 > 1$, however, does lead to new effects as first discussed in [170, 101, 102] and as summarized in the next section.

8.5 Topological inflation

If $16\pi G\eta^2 > 1$, the gravitational forces within the wall are stronger than the forces associated with the self-interaction of the scalar field. This can be seen by the following heuristic argument [170].

Consider the Z_2 model with the quartic potential

$$V(\phi) = \frac{\lambda}{4}(\phi^2 - \eta^2)^2 \tag{8.25}$$

The thickness of the domain wall can be estimated by equating gradient and potential energies, which also agrees with the Bogomolnyi equation (see Eq. (1.31)), in the case when gravitational effects are ignored. The field ϕ gets an expectation value η and so, in the interior of the domain wall,

$$\frac{1}{2}(\nabla\phi)^2 = \frac{\eta^2}{2\delta^2} \sim V(0) = \frac{\lambda}{4}\eta^4 \tag{8.26}$$

and the thickness, δ, is

$$\delta \sim \sqrt{\frac{2}{\lambda\eta}} \tag{8.27}$$

This is an estimate of the length scale on which the scalar field interactions are working.

Next, the length scale associated with gravitational effects is found from the Friedman-Robertson-Walker equation, which relates the space-time expansion rate, H, to the energy density

$$H^2 \sim \frac{8\pi G}{3}\rho \tag{8.28}$$

which, when used inside the wall with $\rho \sim \lambda\eta^4/2$, gives

$$H^{-1} \sim \sqrt{\frac{3}{4\pi G\lambda}\frac{1}{\eta^2}} \tag{8.29}$$

Hence scalar field forces dominate over gravitational forces inside the domain wall if $H^{-1} > \delta$, or the order of magnitude condition,

$$16\pi G\eta^2 < 1 \tag{8.30}$$

Therefore when $16\pi G\eta^2 < 1$ we can expect that gravitational effects are small in the interior of the domain wall. If, however, $H^{-1} < \delta$, the field is approximately smooth over a region where gravitational effects are strong. The field inside the domain wall has potential energy $\sim \lambda \eta^4$ and this is what drives the gravitational effects. Therefore, we expect that the space-time inside the domain wall inflates in the direction normal to the wall, in addition to the inflation parallel to the wall that we have already seen in the thin-wall case (Eq. (8.14)). Furthermore, the field inside the wall is stuck on top of the potential owing to the topology that led to the existence of the wall. So the inflation goes on forever for topological reasons. Hence this inflation is called "topological inflation."

This picture has been confirmed by numerical solution of the coupled scalar field and Einstein equations in [131, 31, 94] with the conclusion that topological inflation inside the Z_2 domain wall occurs for $\eta > 0.33 m_P$ where m_P is the Planck mass defined by $G = 1/m_P^2$.

Domain walls that are undergoing topological inflation cannot however form in the usual way during a cosmological phase transition as we discuss in Section 8.9 below.

8.6 Spherical domain wall

The metric of a thin spherical domain wall has been discussed in the thin-wall limit in [80]. Inside the wall the metric is flat using Birkhoff's theorem (e.g. see [177])

$$ds^2 = dT^2 - dr^2 - r^2(d\theta^2 + \sin^2\theta d\phi^2), \quad r < R(t) \tag{8.31}$$

where $R(t)$ is the radius of the spherical wall and

$$\dot{T} = (1 + \dot{R}^2)^{1/2} \tag{8.32}$$

with overdots denoting derivatives with respect to the proper time of an observer moving with the domain wall. The proper time is related to the time coordinate t via the relation

$$\left(1 - \frac{2GM}{R}\right)\dot{t} = \left(1 - \frac{2GM}{R} + \dot{R}^2\right)^{1/2} \tag{8.33}$$

Outside the sphere, the metric is Schwarzschild with mass parameter M

$$ds^2 = \left(1 - \frac{2GM}{r}\right)dt^2 - \left(1 - \frac{2GM}{r}\right)^{-1} dr^2 - r^2(d\theta^2 + \sin^2\theta d\phi^2), \quad r > R(t) \tag{8.34}$$

The mass is related to the maximum radius of the spherical wall, $R_{\rm m}$, by

$$M = 4\pi\sigma R_{\rm m}^2(1 - 2\pi G\sigma R_{\rm m}) \tag{8.35}$$

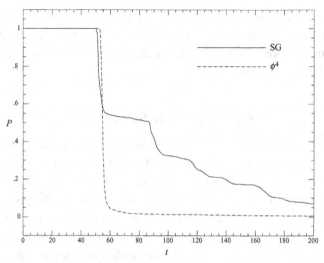

Figure 8.1   Energy in a large volume enclosing a collapsing spherical domain wall in the sine-Gordon and Z_2 models [183] as a function of time. The energy is roughly conserved until the radius becomes comparable to the wall thickness and then decreases sharply. The step-like features in the sine-Gordon models occur because the sphere bounces several times before annihilating. The spherical wall in the Z_2 model annihilates without bouncing. [Figure reprinted from [183].]

provided $R_m < 1/4\pi G\sigma$. If $R_m > 1/4\pi G\sigma$, it means that the spherical domain wall is a black hole even at the maximum value of its radius and the analysis breaks down.

8.7 Scalar and gravitational radiation from domain walls

A collapsing spherical domain wall emits scalar radiation and loses energy. It may be possible to extend the formalism in Section 3.5 to calculate this energy loss. However, such an analysis is not currently available. Instead the energy emission rate has been found numerically and is shown in Fig. 8.1 for spherical walls in the sine-Gordon and Z_2 models [183].

A collapsing spherical domain wall does not emit gravitational radiation since the spherical symmetry implies a vanishing quadrupole moment of the energy-momentum distribution. However, colliding domain walls can lead to gravitational [157] and scalar radiation [175]. A dimensional analysis based on the quadrupole approximation for the gravitational power emitted when two relativistic spherical walls collide gives [157]

$$P_g \sim \frac{GM_B^2}{R^2} \tag{8.36}$$

where $M_B \sim 4\pi\sigma R^2$ is the mass of the bubble and R is the radius upon collision. Numerical analyses of bubble collisions (during first-order phase transitions) found that the quadrupole approximation overestimates the power radiated in gravitational radiation by about a factor of 50 [91].

8.8 Collapse into black holes

If the radius, $R(t)$ of a collapsing spherical domain wall remains larger than the Schwarzschild radius, $R_S = 2GM$, where M is the mass of the domain wall, then the domain wall does not become a black hole. As the wall collapses, it emits scalar radiation and, if this is rapid enough, M decreases sufficiently rapidly so that $R_S < R$ at all times. Whether this happens can be checked explicitly by numerical evolution of the scalar field plus Einstein equations. We expect that if the Schwarzschild radius of the spherical domain wall is smaller than the width of the wall, black holes are not formed since rapid wall annihilation and radiation precede collapse to within the Schwarzschild radius.

The converse case of black hole formation in the case when the scalar radiation is not too rapid is harder to demonstrate convincingly. The reason is that the time evolution of the fields gets slower as the black hole event horizon is about to form. By simply evolving the fields, it is impossible to see the formation of the event horizon and hence conclude that the domain wall collapses to form a black hole. However, it is hard to imagine any other outcome, especially since the scalar radiation rate only becomes significant once the spherical domain wall collapses to a size comparable to the thickness of the wall.

The collapse of a slightly perturbed spherical domain wall has been studied numerically in [182] with the result that the amplitude of perturbations stays constant during the collapse. This means that the ratio of the perturbation amplitude to the radius grows during collapse as $1/R(t)$ and the shape of the wall deviates increasingly from being spherical.

8.9 Cosmological domain walls: formation

The formation of domain walls in a phase transition in flat non-expanding space-time has been discussed in Chapter 6. Since the universe is expanding and cooling, cosmic phase transitions can occur, just as in the laboratory, and domain walls can also form. If the phase transition proceeds quickly on cosmological time scales, the structure of these domain wall networks is similar to those formed in the laboratory and described in Chapter 6. The network is dominated by one infinite domain wall with very complicated topology. However, if the phase transition occurs slowly on cosmological time scales, the expansion can prevent the phase transition from completion. For example, in a first-order phase transition, if the bubble nucleation rate is very slow, the bubbles will not be able to percolate because the expansion increases the separation of the bubbles that have already nucleated. These considerations are important for inflationary cosmology but here we will assume that the phase transition completes since otherwise domain walls would not be formed.

In a model with $16\pi G\eta^2 > 1$ (η is the vacuum expectation value of the scalar field), domain wall formation requires some new considerations [22, 166]. The reason is that the energy density inside the domain walls is larger than that outside. Hence if such inflating domain walls (see Section 8.5) were to form, the space-time expansion rate within them would be greater than that of the ambient cosmological expansion rate in which they were created. It is possible to show that a faster expanding region within a horizon of a slower expanding region can be created only if the null energy condition[1] is violated. The formation of defects proceeds according to the classical dynamics of a scalar field with energy-momentum tensor given by Eq. (8.1). Contracting the energy-momentum tensor twice with a null vector, N^μ, and using $g_{\mu\nu}N^\mu N^\nu = 0$ gives

$$N^\mu N^\nu T_{\mu\nu} = (N^\mu \partial_\mu \phi)^2 \geq 0 \qquad (8.37)$$

Hence the null energy condition is satisfied during defect formation and there is an obstruction to the formation of topologically inflating domain walls. The exception is if the faster expanding region has an extent that is larger than the cosmological horizon. In this situation, the domain wall is fatter than the horizon during the phase transition. Then the particle interaction rate is also slower than the Hubble expansion rate and the particles are not in a thermal state unless they were set up in that state as an initial condition at the Big Bang. The domain wall network that is produced will depend on the initial state of the particles.

8.10 Cosmological domain walls: evolution

If we assume that there is a dense network of walls within our cosmological horizon and that the network does not lose a significant amount of energy to radiation, we can work out the expansion rate of the universe and the scaling of the density of walls.

The energy-momentum of the scalar field that forms the domain walls is given in Eq. (8.1). If we denote an average over a large volume by $\langle \cdot \rangle$ we have

$$\langle T_{\mu\nu} \rangle = \langle \partial_\mu \phi \partial_\nu \phi \rangle - \left\langle g_{\mu\nu} \left\{ \frac{1}{2}(\partial_\alpha \phi)^2 - V(\phi) \right\} \right\rangle \qquad (8.38)$$

---

[1] The null energy condition is $N^\mu N^\nu T_{\mu\nu} > 0$ where N^μ is any null vector and $T_{\mu\nu}$ is the energy-momentum tensor. For fluids with energy density ρ and isotropic pressure p, the null energy condition is $\rho + p > 0$.

We assume that $g_{\mu\nu}$ is a background metric and only dependent on time. Also the field distribution is assumed to be isotropic so that

$$\langle(\partial_x\phi)^2\rangle = \langle(\partial_y\phi)^2\rangle = \langle(\partial_z\phi)^2\rangle \tag{8.39}$$

and

$$\langle\partial_i\phi\partial_j\phi\rangle = 0, \quad i \neq j \tag{8.40}$$

Define

$$\langle\phi'^2\rangle = \frac{1}{3}\langle(\partial_x\phi)^2 + (\partial_y\phi)^2 + (\partial_z\phi)^2\rangle \tag{8.41}$$

If we further assume that the field is dominantly in the form of domain walls that satisfy Eq. (1.31) to a good approximation, we get

$$\langle\phi'^2\rangle = \frac{2}{3}\langle V\rangle \tag{8.42}$$

which leads to

$$\langle T_{xx}\rangle = \frac{5}{6}\langle\dot{\phi}^2\rangle - \frac{2}{3}\langle T_{tt}\rangle \tag{8.43}$$

For slowly varying fields this leads to the effective equation of state $p = -2\rho/3$ where p is the (isotropic) pressure and ρ the energy density [186]. If we assume that the time dependence of ϕ is only due to a boost of the domain walls, we can use $\dot{\phi} = v\gamma\partial_X\phi = v\gamma\sqrt{2V(\phi)}$ and $\partial_x\phi = \gamma\partial_X\phi$, where $X = \gamma(x - vt)$ and γ is the Lorentz factor (see Eq. (1.10)). This leads to

$$\langle T_{xx}\rangle = \left(\langle v^2\rangle - \frac{2}{3}\right)\langle T_{tt}\rangle \tag{8.44}$$

Following Appendix F and treating the wall network as a fluid with equation of state $p = -2\rho/3$, we can write down the solutions for the scale factor and the scaling of the energy density in walls. If the initial conditions are such that the wall density is ρ_0 when the scale factor is a_0, the solution is

$$\rho_{\text{walls}}(a) = \rho_0\frac{a_0}{a}, \quad a(t) = a_0\left(\frac{t}{t_0}\right)^2 \tag{8.45}$$

Note that this derivation ignores processes by which the wall network could lose energy into scalar and gravitational radiation. In addition, the walls interact with surrounding matter and experience friction. These effects make the problem of understanding the evolution of the wall network much more challenging. We now describe some numerical [125, 36, 97, 59] and analytical [77, 78] efforts to understand the evolution of the network.

8.11 Evolution: numerical results

There are two numerical schemes for evolving a network of domain walls. The first is to use the zero thickness approximation for walls. In this approximation, it is hard to treat the collision of walls and the loss of energy from the network into radiation. The second approach is to solve the field theory equations of motion. In this approach, all the degrees of freedom of the system are retained. In fact, a lot of degrees of freedom that are evolved are inessential to the domain wall network and this additional baggage slows down the simulations. In an expanding universe the problem is even more severe because the overall length scales grow larger with time while the domain wall thickness remains the same. Thus the simulation needs to handle very disparate length scales.

In [125, 36, 97, 59], the authors get around these problems by solving the field theory equations of motion but by letting the domain walls expand with the universe.

More specifically, consider the Z_2 model in an expanding space-time with metric

$$g_{\mu\nu} = a^2(\tau)\eta_{\mu\nu} \tag{8.46}$$

where $\eta_{\mu\nu} = \mathrm{diag}(1, -1, -1, -1)$ and τ is the conformal time. The equation of motion is

$$\partial_\tau^2\phi + 2\frac{\dot{a}}{a}\partial_\tau\phi - \nabla^2\phi + \lambda(\phi^2 - a^2\eta^2)\phi = 0 \tag{8.47}$$

In the approach pioneered in [125] the $a^2\eta^2$ in the last term is replaced by a constant, effectively decreasing the vacuum expectation value, η, with Hubble expansion. Since the width of the domain wall is proportional to $1/\eta$, this amounts to letting the thickness of the walls grow in proportion to the scale factor.

The result of this numerical study shows that the areal density, $\mathcal{A}$, i.e. area of walls in a given region divided by the volume of the region, scales inversely as the first power of conformal time

$$\mathcal{A} = \mathcal{A}_0 \left(\frac{\tau_0}{\tau}\right)^p, \quad p \approx 1 \tag{8.48}$$

where the subscript 0 refers to some initial time. This result holds in Minkowski space-time $(a \propto \tau^0)$, radiation-dominated $(a \propto \tau^{1/2})$, and matter-dominated $(a \propto \tau^{1/3})$ cosmologies.

The domain wall network has also been studied by a combination of numerical and analytical techniques that use scaling arguments [13, 14].

8.12 Evolution: analytical work

An analytic technique to study the evolution of non-relativistic interfaces in the condensed matter context was developed in [116] (also see [24, 65, 111]). The technique has been extended to relativistic systems in [77, 78] and we now summarize the main features of this analysis.

The starting point is to define a fictitious scalar field $u(x^\mu)$ such that it vanishes on the domain wall network

$$u(X^\mu(\sigma^a)) = 0, \quad a = 0, 1, 2 \tag{8.49}$$

where the domain wall network is located at $X^\mu(\sigma^a)$ and σ^a denote world-volume coordinates. While $u(x^\mu)$ could have been taken to be the scalar field in the original field theory (say for the Z_2 model), this is not suitable since u is later assumed to be a random field with a Gaussian distribution. The next step is to derive an equation of motion for u.

We define the domain wall world-volume metric as in Eq. (7.16)

$$h_{ab} = g_{\mu\nu}(X)\partial_a X^\mu \partial_b X^\nu \tag{8.50}$$

where $g_{\mu\nu}$ is the ambient space-time metric and the indices a, b refer to world-volume coordinates. Two derivatives of Eq. (8.49) lead to

$$\frac{1}{\sqrt{|h|}}\partial_a(\sqrt{|h|}h^{ab}\partial_b X^\mu)\partial_\mu u + h^{ab}\partial_a X^\mu \partial_b X^\nu \partial_\mu \partial_\nu u = 0 \tag{8.51}$$

As long as the thin-wall limit is valid and, in particular, walls do not intersect, X^μ satisfies the Nambu-Goto equation. When walls do intersect, the Nambu-Goto formalism breaks down. The formalism can continue to be valid provided we impose additional boundary conditions by hand at the intersection point. Depending on the boundary conditions that one imposes at the intersection point, the Nambu-Goto equation can describe intercommuting walls or walls that pass through each other. In the present formalism, the boundary conditions automatically arise from the evolution of the u field. The dynamics of the u field are such that they always describe walls that intercommute [77]. Using the Nambu-Goto equations of motion, Eq. (7.21), then leads to the equation of motion for the fictitious field u

$$[(\partial u)^2 g^{\mu\nu} - \partial^\mu u \partial^\nu u](\partial_\mu \partial_\nu u - \Gamma^\rho_{\mu\nu}\partial_\rho u) = 0 \tag{8.52}$$

where $\Gamma^\rho_{\mu\nu}$ is the Christoffel symbol defined in Eq. (7.22).

To solve Eq. (8.52) we must find a way to handle the non-linear terms. The key point now is that the domain wall network contains a random distribution of walls and hence u is a statistical field. One approach to treat the non-linear terms is to use the mean field approximation. In this approach non-linear terms are replaced by averages of non-linear terms multiplied by a single power of u. For example

$$u^3 \rightarrow \langle u^2 \rangle u \tag{8.53}$$

Further, the distribution of u is assumed to be Gaussian.

After defining the correlators that enter the mean field theory version of Eq. (8.52), the field u satisfies the equation of motion

$$\partial_\tau^2 u + \frac{\mu(\tau)}{\tau}\partial_\tau u - v^2\nabla^2 u = 0 \tag{8.54}$$

where, as in the previous section, τ is the conformal time, The functions μ and v^2 are defined in terms of the assumed two-point correlation functions of u (for details see [77, 78]). Once the solution for u is obtained from the linear differential equation, Eq. (8.54), the average areal density, $\mathcal{A}$, and other quantities may be calculated. The results agree with the scaling in Eq. (8.48).

8.13 Cosmological constraints

The cosmological constraint on domain walls is remarkably robust, being almost independent of the field theory, details of the phase transition, and cosmology [186]. At any time after the domain wall forming phase transition, the vacua in different cosmological horizons are uncorrelated. This means that there is at least one domain wall per horizon. The minimum area of a horizon size domain wall is $\sim H^{-2}$ where H^{-1} is the horizon size. Therefore the domain wall energy density averaged over a horizon volume is $\rho_{\mathrm{walls}} \sim \sigma H$. Comparing this to the critical density of the universe,[2] we get

$$\Omega_{\mathrm{walls}} \equiv \frac{\rho_{\mathrm{walls}}}{\rho_c} \sim \frac{G\sigma}{H} \sim G\sigma t \tag{8.55}$$

where t is the cosmic time. (We have taken $H \sim 1/t$ which is true in a Friedman-Robertson-Walker cosmology in which $a(t) \propto t^\alpha$ with $0 < \alpha < 1$.) Hence, as time proceeds, there comes an epoch when the domain walls are the dominant form of energy in the universe. This happens at time t_* given by

$$t_* \sim \frac{1}{G\sigma} \tag{8.56}$$

Now $\sigma \sim \eta^3$ (e.g. Eq. (1.20)) up to factors of coupling constants which we assume are order unity. We also know particle physics fairly well up to an energy scale of about 100 GeV (approximately the electroweak scale) and have not seen any scalar fields yet. So the minimum value of σ is about $(100\,\mathrm{GeV})^3$. Walls of this tension would have started dominating the universe at (see Appendix A for numerical values)

$$t_*\bigg|_{\mathrm{min}} \sim \frac{m_{\mathrm{P}}^2}{\eta^3} \sim 10^8 \text{ s} \tag{8.57}$$

---

[2] The critical density of the universe is defined as $\rho_c = 3H^2/8\pi G$, where $H(t) = \dot{a}/a$ is the Hubble expansion rate defined in terms of the scale factor $a(t)$ and its time derivative, $\dot{a}$.

or approximately 10 years after the Big Bang. Once the walls dominate, the universal expansion becomes $a \propto t^2$ (Eq. (F.5)). This is unacceptable for several reasons. For example, since the domain wall dominated universe accelerates ($\ddot{a} > 0$), density perturbations that are larger than the horizon keep getting stretched and stay larger than the horizon. This means that super-horizon density perturbations can never re-enter the horizon, which is an essential condition for them to start growing to form the galaxies, clusters, and large-scale structures that we currently observe. Even the growth of sub-horizon density perturbations is suppressed owing to cosmic acceleration.

A second constraint on a network of cosmic domain walls acting as a fluid with equation of state $p = -2\rho/3$ comes from the measured expansion rate of the universe using supernovae data [127, 118]. These surveys find that the equation of state parameter, $w \equiv p/\rho$, for our universe is less than about -0.8 [11, 128]. However, a universe dominated by a network of static ("frustrated") domain walls [25] would have $w \approx -0.67$.

Another possibility that has been considered is that perhaps there are some features that are missing in the standard model of particle physics, and that there indeed are very light domain walls in the universe [76]. Such light walls, if light enough, would be benign and could potentially play a role in cosmology. If we require that the domain walls not dominate the universe until the present time ($\sim 10^{17}$ s), Eq. (8.57), gives $\eta < 100$ MeV. Other cosmological constraints, such as arising from the isotropy of the cosmic microwave background can be used to put similar or somewhat stronger bounds on η [144, 158].

8.14 Constraints on and implications for particle physics

Let us summarize the picture that has emerged in this chapter.

- If a field theory has discrete symmetries that are spontaneously broken in the ground state, it must contain domain wall solutions.
- If high-energy particle physics is described by such a field theory and the discrete symmetry gets spontaneously broken in the early universe, cosmic domain walls are produced.
- If the standard model is complete at energies below 100 GeV, then there can be no domain walls in the universe and no spontaneously broken discrete symmetries in particle physics.

A closer examination of this sequence of arguments reveals a few loopholes that allow for spontaneously broken discrete symmetries in particle physics. First, there is the possibility that the discrete symmetry was broken right from the moment of the Big Bang. Then the whole universe could have been in one of the many discrete vacua at its very creation and no domain walls would be formed even though the

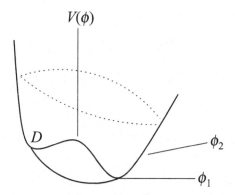

Figure 8.2   Sketch of the potential in the model of Eq. (8.58). Domain walls arise owing to a 2π change in the angular field variable and the location of the field inside the wall is marked by D. Such domain walls can terminate on strings, and the field within the string is located at $\phi = 0$.

underlying particle-physics theory could have broken discrete symmetries. This kind of scenario has been studied for magnetic monopoles in [50]. A related possibility is that, if the universe went through a period of superluminal expansion ("inflation"),[3] then correlations extend to scales that are vastly larger than our current horizon and our region of the universe is very likely to be free of any domain walls [97]. In another variant, domain walls are formed but subsequently inflated away.

All the above loopholes only apply to very high-energy domain walls where quantum gravity and/or inflation effects are relevant. If a particle-physics model has spontaneously broken discrete symmetries at lower energy scales (but still larger than ~ 100 MeV) no loopholes are known and the model is ruled out based on the "cosmological domain wall catastrophe." However, there is still the possibility that metastable or biased domain walls (see Section 6.8) can exist for some time in the universe. We now describe these two possibilities.

8.15 Metastable domain walls

In certain field theories, it is possible for domain walls to get punctured. To see how this can happen, consider the potential for a *complex* scalar field ϕ

$$V(\phi) = \frac{\lambda}{4}(|\phi|^2 - \eta^2)^2 - \frac{\alpha\eta}{32}(\phi + \phi^*)^3 \tag{8.58}$$

where we assume $0 < \alpha << \lambda$. The shape of this potential is shown in Fig. 8.2. The first term is minimized when $|\phi| = \eta$ and, restricting ϕ to the submanifold $|\phi| = \eta$,

---

[3] Inflation occurs when the universe is dominated by a field that has an equation of state with $-\rho < p < -\rho/3$. Then the expansion rate of the universe is superluminal and volumes that are larger than the horizon can get correlated.

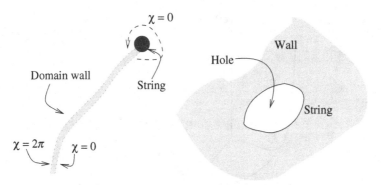

Figure 8.3   Cross-section of a wall that terminates on a string is shown on the left, and a wall with a puncture bordered by a string is shown on the right.

the second term is minimized when $\phi + \phi^* = +2\eta$. Another way of writing the potential is by setting $\phi = \psi \exp(i\chi)$ and then ψ, χ are real fields. Then

$$V(\phi) = \frac{\lambda}{4}(\psi^2 - \eta^2)^2 - \frac{\alpha \eta}{4}\psi^3 \cos^3 \chi \qquad (8.59)$$

The extrema of V are at $\psi = 0$ and at

$$\psi = \eta \left[\frac{3\alpha + \sqrt{9\alpha^2 + 64\lambda^2}}{8\lambda} \right], \qquad \chi = n\pi \qquad (8.60)$$

where n is an integer. The true vacua occur when n is an even integer. For example, domain wall solutions exist with the boundary condition $\chi(x = -\infty) = 0$, $\chi(x = +\infty) = 2\pi$.

Now consider a domain wall in the model in $3 + 1$ dimensions. Such a domain wall can terminate as shown in Fig. 8.3 since the path from $\chi = 0$ to $\chi = 2\pi$ can be contracted by lifting it over the top ($\psi = 0$) of the potential. While we have not described cosmic strings here, in models such as Eq. (8.58), we can have finite sections of open walls that are bordered by strings. Walls can also get punctured by holes that are bounded by strings. For further discussion of walls bounded by strings, we refer the reader to [88, 168, 171].

The evolution of a network of walls that can have punctures is very different from that of stable walls because a puncture can grow and eat up the wall. This provides a very efficient way for the wall network to lose energy and so the network never dominates the universe [168].

Another scheme that allows for the universe to have a finite period of time with domain walls is if a discrete symmetry is broken and then restored (see Section 6.1). Walls would be formed at the first phase transition and then they would dissolve at the second phase transition when the symmetry is restored. However, this scheme would imply an unbroken discrete symmetry in the low-energy particle physics. We do not know of such a discrete symmetry although the possibility cannot be excluded.

Finally, domain walls could have existed for some time in the early universe if there is an approximate discrete symmetry in the high-energy particle-physics model [60]. We have already seen an example of an approximate discrete symmetry in the $SU(5)$ Grand Unification model discussed in Section 2.1. If the cubic coupling in the potential in Eq. (2.5) is small, it can be ignored and the resulting model has $SU(5) \times Z_2$ symmetry, with all the domain wall solutions discussed in Section 2.2. A simpler example is that of the $\lambda\phi^4$ together with a small cubic term. The potential is

$$V(\phi) = -\frac{m^2}{2}\phi^2 + \gamma m\phi^3 + \frac{\lambda}{4}\phi^4 \tag{8.61}$$

Now the model still has two local minima but they are not exactly degenerate if γ is very small (see Fig. 6.11). At the phase transition, a network of domain walls is formed and the typical separation and curvature scale of the domain walls is given by the correlation length ξ_0. With time the curvature scale grows and is denoted by $R(t)$. So the force per unit area on the wall owing to tension is $\sim \sigma/R(t)$ where σ is the energy density of the wall. There is also a pressure difference pushing the wall toward the vacuum with the lowest energy. This pressure is given by the energy difference between the vacua and hence is proportional to γ

$$p \sim \gamma m\eta^3 \tag{8.62}$$

where η is the vacuum expectation value of the field. Therefore the tension is much larger than the pressure, and the dynamics of the wall network are unaffected by the pressure difference coming from the cubic term as long as

$$R(t) < \frac{\sigma}{\gamma m\eta^3} \sim \frac{1}{\gamma\eta} \tag{8.63}$$

Once the network has evolved to a point where this condition is not met, the pressure becomes important and drives the domain wall network such that the whole system reaches the true vacuum. From the area scaling law in Eq. (8.48), it follows that $R(t)$ grows linearly with conformal time.[4] Therefore, in a radiation dominated universe,

$$R = R_0 \frac{\tau}{\tau_0} = R_0 \left(\frac{t}{t_0}\right)^{1/2} \tag{8.64}$$

---

[4] The scaling law holds at late times after friction becomes unimportant. At earlier times, R grows as a different power of conformal time [87].

where R_0 can be taken to be the correlation length at the time of the phase transition t_0. Inserting this relation in Eq. (8.63) and using $R_0 \sim 1/\eta$, we get that the walls survive for a duration

$$t_{\text{walls}} \sim \frac{t_0}{\gamma^2} \tag{8.65}$$

If γ is small, the walls can survive for many Hubble expansions. In fact, if the walls survive for a long time, they might start dominating the density of the universe before they disappear.

Even if domain walls are present in the universe for a relatively short time, they can still have important implications for cosmology. As the wall network evolves, the ambient matter interacts with the walls. Magnetic monopoles can get trapped on domain walls, leading to faster annihilation. This is the sweeping scenario discussed in [52]. In addition, the eventual collapse of domain walls can lead to black hole formation. These issues have received some attention but have yet to be studied in detail.

8.16 Open questions

1. What happens when a black hole collides with a domain wall? Does it get stuck on the wall? Or does it pass through? For a discussion from the gravitational point of view, see [29, 148].
2. Develop an analytical formulation (perhaps along the lines of Section 3.5) to calculate the scalar radiation rate from collapsing domain walls.
3. Discuss the cosmology of superconducting domain walls.
4. What is the outcome of the $SU(5)$ Grand Unified phase transition when the cubic coupling is small? Are domain walls formed? How does the network evolve?

9

Kinks in the laboratory

In this chapter we discuss two laboratory systems where kinks are known to exist. The first system is *trans* polyacetylene which has a broken Z_2 symmetry as in the $\lambda\phi^4$ model. The second system is a Josephson junction transmission line, which is a laboratory realization of the sine-Gordon system. Helium-3 is another laboratory system that contains a wide variety of topological defects and the reader is referred to [174] for a discussion. In the third section of this chapter we describe Scott Russell's solitons in water. These solitons are not topological like the others discussed in this book but we include the discussion anyway since the reader's curiosity may have been aroused by the story in the Preface.

9.1 Polyacetylene

Polyacetylene consists of a linear chain of CH bonds. A sequence of x units is written as $(CH)_x$. In the ground state of polyacetylene, the carbon atom forms three σ bonds, one of them is to the H in the CH unit, one to the unit on the left and one to the right. In addition, there is one more electron orbital that can cause bonding. This is called the π electron, and the π bond can form to the left or to the right. Then there are two possible sequences – first when the double (σ and π) bond is to the carbon on the right and the single to the left, the second when the double bond is to the left and the single to the right. These two possibilities are illustrated in Fig. 9.1 [149] in the *trans* configuration of polyacetylene.[1]

The average bond length $a \approx 1.22$ Å but the CH units are displaced so as to make double bonds (slightly) shorter than the single bonds. The physical displacements u_n along the horizontal axis in the two structures are depicted in Fig. 9.1. Qualitatively, the essential point is that the π electrons have to choose to either form the double

---

[1] In the *cis* configuration, there are also two states related by the left-right transformation but they are not degenerate in energy.

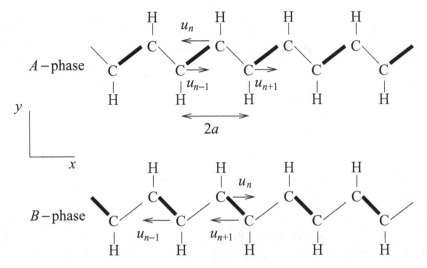

Figure 9.1   Structure of the two degenerate ground states of *trans* polyacetylene. The upper structure is denoted by A and the lower by B. Double bonds are denoted by heavy lines.

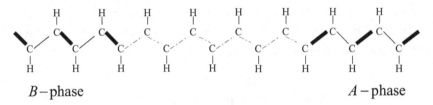

Figure 9.2   If the B state occurs on the left side of a chain and the A on the right, there is a kink in between where the simple alternating structure cannot be maintained. The kink is the region where the alternate single-double bonds do not exist.

bond to the left or to the right. Hence there is a Z_2 symmetry which is broken in the ("dimerized") ground state. Kinks form if different ground states are chosen at different locations (Fig. 9.2). The center of the kink is located at the CH unit where the π electron wavefunction is equally shared between the CH units to the left and right.

The Hamiltonian of the system depends on the displacement variables, u_n and on the locations of the π electrons

$$H = -\sum_{n,s}(t_{n+1,n}c_{n+1,s}^{\dagger}c_{n,s} + h.c.) + \sum_{n}\frac{K}{2}(u_{n+1} - u_n)^2 + \sum_{n}\frac{M}{2}\dot{u}_n^2 \qquad (9.1)$$

where

$$t_{n+1,n} = t_0 - \alpha(u_{n+1} - u_n) \qquad (9.2)$$

is the hopping integral to leading order in displacements. The operators $c_{n,s}^\dagger$ and $c_{n,s}$ are creation and annihilation operators for electrons of spin s on the nth CH group. The parameter K is the effective spring constant of the σ bonds and M is the mass of the CH group.

To connect with the discussion of Chapter 1, the displacement variable

$$\phi_n = (-1)^n u_n \tag{9.3}$$

can be viewed as a scalar field defined on a lattice interacting with a fermion (the π electron). The last two terms in Eq. (9.1) correspond to gradient and time derivative terms of a continuum field $\phi(x)$ that corresponds to the discrete variables, ϕ_n. The first term describes interactions between ϕ and the electrons. The effective interaction for the ϕ field, after integrating out the fermionic variables, must respect the Z_2 symmetry, and hence corresponds to a ϕ^4 interaction to lowest order. Therefore a non-relativistic version of the Z_2 model of Eq. (1.2) captures some of the gross features of polyacetylene.

The properties of kinks in polyacetylene (Fig. 9.2) have been studied in [150] using the Hamiltonian in Eq. (9.1) with the result that the kink width is approximately 14 lattice spacings and the mass is approximately six electron masses [152] in good agreement with experiments [75].

The quantum properties of polyacetylene kinks have also been studied. In Section 5.3 we discussed how kinks can carry fractional quantum numbers [83]. Polyacetylene kinks also carry fractional quantum numbers and electric charge, namely "half a bond" or $\pm(2e)/2$ charge since each bond consists of two electrons (one from each atom at either end of a bond) [150]. Indeed, in a chain where two single bonds are followed by a double bond (instead of the alternating single and double bonds in *trans* polyacetylene) the fractional charge can be shown to be one-third of a bond [151]. Reference [68] generalizes these ideas much further and shows that solitons may even carry irrational charges.

9.2 Josephson junction transmission line

We follow [134] in deriving the sine-Gordon equation for the Josephson transmission line.

Let us recall the basics of a transmission line, schematically shown in Fig. 9.3 [54]. A potential difference is applied to the ends of two elements of a transmission line e.g. the two cables of a coaxial cable. The potential, V, and current, I, in each of the wires are functions of the location on the transmission line, namely the x coordinate, and also of time. There is also a potential difference between the wires, and the current in the two wires can be different, but this is not shown in

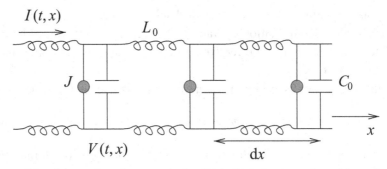

Figure 9.3   Schematics of a transmission line with inductance L_0 and capacitance C_0 per unit length. The symbols marked J represent a second coupling between the two transmission components. This coupling is absent in an ordinary transmission line but represents the tunneling current in the Josephson transmission line.

the figure. We will only be considering the potential and current distribution along a single wire. Let L_0 denote the inductance per unit length of the line, and C_0 the capacitance per unit length. Then Faraday's law of induction tells us that the induced e.m.f. between points $x + dx$ and x is proportional to the rate of change of the current in the segment within those points

$$V(t, x + dx) - V(t, x) = -(L_0 dx)\frac{\partial I}{\partial t} \tag{9.4}$$

or

$$\frac{\partial V}{\partial x} = -L_0 \frac{\partial I}{\partial t} \tag{9.5}$$

Charge accumulates on the segment from x to $x + dx$ in time dt owing to the different entering and exiting currents. The charge on the segment is also given by the capacitance times the potential. Hence

$$I(t, x + dx) - I(t, x) = -(C_0 dx)\frac{\partial V}{\partial t} \tag{9.6}$$

or

$$\frac{\partial I}{\partial x} = -C_0 \frac{\partial V}{\partial t} \tag{9.7}$$

Equations (9.5) and (9.7) can be combined to obtain wave equations for the current and the potential.

   A Josephson junction transmission line differs from the ordinary transmission line described above in that the two "wires" are superconductors and they are separated by a thin insulator. This set-up is shown in Fig. 9.4. Current can tunnel through the insulator and jump from one wire to the other. Hence the charge on a

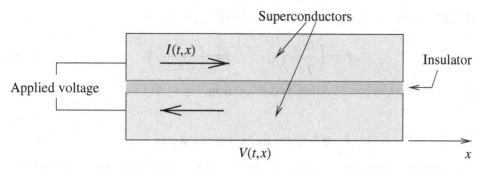

Figure 9.4   A Josephson junction transmission line is constructed by separating two superconducting plates by a thin layer of an insulating material.

segment also changes owing to the Josephson current and Eq. (9.7) gets modified

$$\frac{\partial I}{\partial x} = -C_0 \frac{\partial V}{\partial t} - j_J(x, t) \tag{9.8}$$

where j_J is the Josephson current per unit length.

The charge carriers (Cooper pairs) in either superconductor are described by macroscopic wavefunctions

$$\psi_1 = \sqrt{\rho_1} e^{i\phi_1}, \qquad \psi_2 = \sqrt{\rho_2} e^{i\phi_2} \tag{9.9}$$

where ρ_1 and ρ_2 are the charge carrier number densities in the superconductors. The tunneling Josephson current per unit area is [55]

$$j_J = j_0 \sin \phi \tag{9.10}$$

where j_0 is the maximum Josephson current and is proportional to $\sqrt{\rho_1 \rho_2}$, and

$$\phi = \phi_1 - \phi_2 \tag{9.11}$$

The Schrödinger equations for ψ_1 and ψ_2 imply

$$\frac{\partial \phi}{\partial t} = \frac{q}{\hbar} V \tag{9.12}$$

where V is the potential difference across the junction and $q = 2e$ is the electric charge of a Cooper pair.

From Eqs. (9.5) and (9.12) we obtain

$$I = \frac{-\hbar}{q L_0} \frac{\partial \phi}{\partial x} \tag{9.13}$$

Now we can insert this expression for I, and V as found from Eq. (9.12), in Eq. (9.8) to get

$$\frac{\partial^2 \phi}{\partial t^2} - \frac{1}{L_0 C_0} \frac{\partial^2 \phi}{\partial x^2} + \frac{j_0 q}{C \hbar} \sin \phi = 0 \tag{9.14}$$

Rescaling t and x by the Josephson time and length scales

$$\tau = \left(\frac{\hbar C_0}{q J_0}\right)^{1/2}, \qquad l = \left(\frac{\hbar}{q J_0 L_0}\right)^{1/2} \tag{9.15}$$

gives the sine-Gordon equation as derived from Eq. (1.51) with $\alpha = 1 = \beta$.

9.3 Solitons in shallow water

The solitons discussed in this book have all had a topological origin. In contrast, the solitons first discovered by Scott Russell in a water channel, and mentioned in the Preface, have their origin in the non-linearities of hydrodynamics and do not have a topological origin.

The first step to show the existence of the water solitons is to derive the Korteweg-deVries (KdV) equation for waves of long wavelength moving in one direction in shallow water. We do not give this derivation here and instead refer the reader to, for example, Section 13.11 of [180].

The KdV equation is

$$\frac{\partial u}{\partial t} + u\frac{\partial u}{\partial x} + \delta^2\frac{\partial^3 u}{\partial x^3} = 0 \tag{9.16}$$

where u is related to the height of the fluid surface and δ is a parameter. The soliton solution is [185]

$$u = u_\infty + (u_0 - u_\infty)\mathrm{sech}^2\left[\frac{X - X_0}{\Delta}\right] \tag{9.17}$$

where $X = x - vt$, X_0 is a constant, v is the velocity of the soliton,

$$\Delta = \delta\left[\frac{u_0 - u_\infty}{12}\right]^{-1/2} \tag{9.18}$$

The velocity of the soliton is given by

$$v = u_\infty + \frac{u_0 - u_\infty}{3} \tag{9.19}$$

in terms of the arbitrary constants u_0 and u_∞. Note that the amplitude of the soliton and the velocity are related.

9.4 Concluding remarks

There are a number of situations where solitons have been discussed in the particle physics literature. Most of these discussions, such as of domain walls in the $SU(5) \times Z_2$ model in Chapter 2, have been in the framework of Grand Unified

Theories. The attention has mostly focused on magnetic monopoles and strings because monopoles seem inevitable in this class of theories and strings are less constrained by cosmology. Similar topological structures also exist in the standard model of electroweak interactions but the monopoles are confined and the strings are unstable [2]. Domain wall and string solutions also exist in QCD in various external conditions, for example in high density matter such as might be present in the interiors of neutron stars [142]. Unlike solitons in the laboratory, however, solitons in particle physics and cosmology have not yet been discovered experimentally. Given the very similar underpinnings of laboratory and particle physics systems, there is hope that this situation will soon change.

9.5 Open questions

1. Is there a condensed matter system with spontaneously broken permutation symmetry? Discuss the domain walls in that system and whether a lattice can exist. Can the walls be observed experimentally?
2. If there are QCD domain walls in neutron stars, how might they be observed from Earth?

Appendix A

Units, numbers and conventions

We will work in natural units in which $\hbar = c = 1$. In these units, all dimensionful quantities have dimensions of mass to some power. One way to convert from mass (g) to length (cm) and time (s), is to remember the values for the Planck mass, time, and length: $m_P = 1.2 \times 10^{19}$ GeV, $t_P = 5.4 \times 10^{-43}$ s, $l_P = 1.6 \times 10^{-33}$ cm. Also, $m_P t_P = 1 = m_P l_P$ in natural units. It is also useful to remember $m_P = 2.2 \times 10^{-5}$ g and, when dealing with magnetic fields, the conversion: 1 Gauss $= 1.95 \times 10^{-20}$ GeV^2. In addition, for cosmological estimates it is convenient to know that 1 pc $= 3.1 \times 10^{18}$ cm.

The metric signature is taken to be $(+, -, -, -)$.

Appendix B

$SU(N)$ generators

$SU(N)$ is the group of special (unit determinant), unitary, $N \times N$ complex matrices.[1] By considering the various constraints on the $2N^2$ real components of the matrix owing to the special and unitary conditions, we can see that the matrix has $N^2 - 1$ independent degrees of freedom. Then, if $g \in SU(N)$, we can write

$$g = \exp(i\alpha_a T^a) \qquad \text{(B.1)}$$

where a sum over $a = 1, \ldots, N^2 - 1$ is implicit, α_a are real constants, and T^a are the "generators" of the group. The T^a satisfy the $SU(N)$ Lie algebra and can be represented by matrices of various dimensions. In the $N = 2$ ($SU(2)$) case, the two-dimensional representation is in terms of Pauli spin matrices, $T^a = \sigma^a/2$, or explicitly

$$T^1 = \frac{1}{2}\begin{pmatrix} 0 & 1 \\ 1 & 0 \end{pmatrix}, \qquad T^2 = \frac{1}{2}\begin{pmatrix} 0 & -i \\ i & 0 \end{pmatrix}, \qquad T^3 = \frac{1}{2}\begin{pmatrix} 1 & 0 \\ 0 & -1 \end{pmatrix} \qquad \text{(B.2)}$$

The Lie algebra is

$$[T^a, T^b] = i\epsilon^{abc} T^c \qquad \text{(B.3)}$$

where ϵ^{abc} is the totally antisymmetric tensor. One can also easily construct the higher dimensional representations. It is conventional to normalize the generators to satisfy

$$\text{Tr}(T^a T^b) = \frac{1}{2}\delta^{ab} \qquad \text{(B.4)}$$

where δ^{ab} is the Kronecker delta.

To get a set of generators for $SU(N)$, it is simplest to build on the $SU(2)$ generators in Eq. (B.2). First, one puts the Pauli spin matrices in the upper left-hand corner and obtains three $SU(N)$ generators

$$T^a = \frac{1}{2}\begin{pmatrix} \sigma^a & 0 & \cdots \\ 0 & 0 & \cdots \end{pmatrix}, \qquad a = 1, 2, 3 \qquad \text{(B.5)}$$

Then one puts the off-diagonal Pauli spin matrices in the off-diagonal positions. Since there are $N(N-1)/2$ off-diagonal positions of which two have already been filled by the $a = 1, 2$ generators, we can construct $N(N-1) - 2$ more generators by filling each

---

[1] For a review of group theory in particle physics, see [62].

remaining position by either 1 (as in σ^1) or by $\pm i$ (as in σ^2). These look like

$$\frac{1}{2}\begin{pmatrix} 0 & 0 & \cdots & \cdots & \cdots \\ 0 & 0 & \cdots & \cdots & \cdots \\ \cdot & & \cdots & 1_{jk} \\ 0 & \cdots & 1_{kj} & & \\ \cdots & \cdots & \cdots & \cdots & \cdots \end{pmatrix}, \qquad \frac{1}{2}\begin{pmatrix} 0 & 0 & \cdots & & \cdots \\ 0 & 0 & \cdots & \cdots & \cdots \\ \cdot & & \cdots & -i_{jk} \\ 0 & \cdots & i_{kj} & & \cdots \\ \cdots & \cdots 0 & \cdots & & \cdots \end{pmatrix} \qquad (B.6)$$

where the subscripts j, k denote the position in the matrix.

Finally we construct the diagonal generators. These are written by putting a series of 1s in say, n, successive diagonal positions, and then entering $-n$ in the nn entry of the matrix. This scheme ensures that the generator is traceless and the resulting matrix is

$$\text{diag}(1, \ldots, 1_n, -n, 0, \ldots, 0) \qquad (B.7)$$

where 1_n denotes 1 in the nn entry. The normalization is then fixed using the convention in Eq. (B.4) to get the generator

$$\frac{1}{\sqrt{2n(n+1)}}\text{diag}(1, \ldots, 1_n, -n, 0, \ldots, 0) \qquad (B.8)$$

In this way we construct $N - 1$ diagonal generators, one for each value of n. The third Pauli matrix is already included as the $a = 3$ generator.

As a check, we find that the total number of generators constructed is $3 + (N(N-1)-2) + (N-2) = N^2 - 1$ and this agrees with the degrees of freedom in $SU(N)$.

In the $SU(5)$ Grand Unified model discussed in Chapter 2 an alternate set of diagonal generators is useful.

$$\lambda_3 = \frac{1}{2}\text{diag}(1, -1, 0, 0, 0)$$

$$\lambda_8 = \frac{1}{2\sqrt{3}}\text{diag}(1, 1, -2, 0, 0)$$

$$\tau_3 = \frac{1}{2}\text{diag}(0, 0, 0, 1, -1)$$

$$Y = \frac{1}{2\sqrt{15}}\text{diag}(2, 2, 2, -3, -3)$$

After the $SU(5)$ symmetry is broken by the canonical vacuum expectation value of Φ (Eq. (2.6)), λ_3 and λ_8 are generators of the unbroken $SU(3)$, τ_3 of $SU(2)$, and Y of $U(1)$.

Appendix C

Solution to a common differential equation

We have often encountered a differential equation of the type

$$-\frac{d^2\psi}{dx^2} + \left[\epsilon - v\cosh 2\mu - v\sinh 2\mu\,\tanh x + v\cosh^2\mu\,\text{sech}^2 x\right]\psi = 0 \qquad \text{(C.1)}$$

where v, μ are parameters and ϵ is the eigenvalue. This differential equation has been solved in Section 12.3 of [113] where the Schrödinger problem has also been extensively studied. Here we reproduce the solution.

The solution is given in terms of new parameters a and b

$$a = \frac{1}{2}\sqrt{ve^{2\mu} - \epsilon} - \frac{1}{2}\sqrt{ve^{-2\mu} - \epsilon} \equiv \frac{1}{2}\kappa_+ - \frac{1}{2}\kappa_- \qquad \text{(C.2)}$$

$$b = \frac{1}{2}\sqrt{ve^{2\mu} - \epsilon} + \frac{1}{2}\sqrt{ve^{-2\mu} - \epsilon} \equiv \frac{1}{2}\kappa_+ + \frac{1}{2}\kappa_- \qquad \text{(C.3)}$$

Then, with

$$\psi = e^{-ax}\text{sech}^b x\, F(x) \qquad \text{(C.4)}$$

the equation for F becomes

$$F'' - 2[a + b\tanh x]F' + [v\cosh^2\mu - b(b+1)]\text{sech}^2 x\, F = 0 \qquad \text{(C.5)}$$

where primes denote derivatives with respect to x. Defining

$$u = \frac{1}{2}[1 - \tanh x] \qquad \text{(C.6)}$$

we get the hypergeometric equation

$$u(1-u)\frac{d^2 F}{du^2} + [a + b + 1 - 2(b+1)u]\frac{dF}{du} + [v\cosh^2\mu - b(b+1)]F = 0 \qquad \text{(C.7)}$$

The general solution may be found in [71]

$$F = AF_1 + BF_2 \qquad \text{(C.8)}$$

where A and B are constants of integration and

$$F_1 = F(\alpha, \beta; \gamma; u) \qquad \text{(C.9)}$$
$$F_2 = u^{1-\gamma} F(\alpha - \gamma + 1, \beta - \gamma + 1; 2 - \gamma; u) \qquad \text{(C.10)}$$

157

where

$$\alpha = b + \frac{1}{2} - \sqrt{v\cosh^2\mu + \frac{1}{4}}$$

$$\beta = b + \frac{1}{2} + \sqrt{v\cosh^2\mu + \frac{1}{4}} \qquad (C.11)$$

$$\gamma = a + b + 1 \qquad (C.12)$$

and γ is assumed to not be an integer.

The general analysis can be taken further by considering the solution at $x = \pm\infty$. A solution that is regular at $x \to \infty$ (i.e. $u = 0$) is obtained by setting $B = 0$ in Eq. (C.8). Regularity at $x = -\infty$ ($u = 1$) is only obtained for certain values of ϵ, and thus the energy levels are quantized. The details of the general analysis may be found in Section 12.3 of [113].

In this book, we have often encountered the special case with $\mu = 0$. Then, bound states are obtained for the following discrete values of $b > 0$

$$b_n = \sqrt{v + \frac{1}{4}} - \left(n + \frac{1}{2}\right) \qquad (C.13)$$

where $n = 0, 1, 2, \ldots, N$ with N determined by $b_{N+1} \leq 0$. The discrete eigenvalues of ϵ follow from the definition in Eq. (C.3)

$$\epsilon_n = (2n + 1)\sqrt{v + \frac{1}{4}} - \left(n^2 + n + \frac{1}{2}\right) \qquad (C.14)$$

Appendix D

Useful operator identities

Identity 1

We wish to prove[1]

$$e^{A+B} = e^A e^B e^{C/2} \qquad (D.1)$$

where $C = [B, A]$ is assumed to commute with A and B.

Let

$$S(x) = e^{(A+B)x} \qquad (D.2)$$

where x is a parameter. Write

$$S(x) = e^{Ax} U(x) \qquad (D.3)$$

where U is an unknown matrix-valued function. Then

$$(A + B)S(x) = \frac{dS}{dx} = Ae^{Ax} U(x) + e^{Ax} \frac{dU}{dx} \qquad (D.4)$$

which leads to

$$\frac{dU}{dx} = e^{-Ax} B e^{+Ax} U(x) \qquad (D.5)$$

Now

$$Be^{Ax} = B \sum_n \frac{(Ax)^n}{n!} = \sum_n \frac{(Ax)^n}{n!} B + \sum_n \frac{[B, A^n]x^n}{n!} \qquad (D.6)$$

Also,

$$[B, A^n] = A^{n-1}[B, A] + [B, A^{n-1}]A = \cdots = [B, A]nA^{n-1} \qquad (D.7)$$

provided $C = [B, A]$ commutes with A. Therefore

$$Be^{Ax} = e^{Ax} B + [B, A]xe^{Ax} \qquad (D.8)$$

and

$$\frac{dU}{dx} = (B + [B, A]x)U(x) \qquad (D.9)$$

---

[1] Several of the proofs in this Appendix were provided by Harsh Mathur, private communication (2005).

This equation can be solved to get

$$U(x) = \exp(Bx + [B, A]x^2/2) \tag{D.10}$$

This solution satisfies the boundary condition $U(0) = 1$.

Hence

$$S(x) = e^{(A+B)x} = e^{Ax}e^{Bx}e^{[B,A]x^2/2} \tag{D.11}$$

With $x = 1$, we get the desired result.

Identity 2

Here we outline a proof of the identity

$$: e^{A+B} :=: e^A :: e^B : e^D \tag{D.12}$$

where $D = [A^+, B^-]$ is assumed to be a c-number. Also, a linear decomposition of A and B is assumed $A = A^+ + A^-$, $B = B^+ + B^-$ and the superscripts $\pm$ refer to terms proportional to creation $(+)$ and annihilation $(-)$ operators. Expressions sandwiched between : : are normal ordered, and so annihilation operators are placed to the right of creation operators.

The first step is the identity

$$: e^A := e^{A^-}e^{A^+} \tag{D.13}$$

This can be proved by explicit expansion of the exponentials.

Then

$$: e^A :: e^B := e^{A^-}e^{A^+}e^{B^-}e^{B^+} \tag{D.14}$$

and

$$: e^{A+B} := e^{A^-}e^{B^-}e^{A^+}e^{B^+} \tag{D.15}$$

since $[A^-, B^-] = 0 = [A^+, B^+]$.

Now we use the identity, Eq. (D.1) proved in the previous section to exchange the order of the middle two factors in Eq. (D.14) and together with Eq. (D.15) gives the identity in Eq. (D.12).

Identity 3

Here we wish to show

$$A : e^B :=: \{A + [A^+, B^-]\}e^B : \tag{D.16}$$

$$A : e^B := \sum_{n=0}^{\infty} \frac{(A^+ + A^-)^n}{n!} \sum_{k=0}^{n} \frac{n!}{k!(n-k)!}(B^-)^k(B^+)^{n-k} \tag{D.17}$$

$$= \sum_{n=0}^{\infty} \frac{1}{n!} \sum_{k=0}^{n} \frac{n!}{k!(n-k)!}\{(B^-)^k A^+ + [A^+, (B^-)^k]$$

$$+ A^-(B^-)^k\} (B^+)^{n-k} \tag{D.18}$$

Now use

$$[A^+, (B^-)^k] = k(B^-)^{k-1}[A^+, B^-] \tag{D.19}$$

provided that $[A^+, B^-]$ is a c-number. Therefore

$$A : e^B : = \sum_{n=0}^{\infty} \frac{1}{n!} \sum_{k=0}^{n} \frac{n!}{k!(n-k)!} \{(B^-)^k A^+ + k(B^-)^{k-1}[A^+, B^-]$$

$$+ A^-(B^-)^k\} (B^+)^{n-k} \tag{D.20}$$

$$= : (A + [A^+, B^-])e^B : \tag{D.21}$$

Similarly we can show

$$: e^B : A =: e^B(A + [B^+, A^-]) : \tag{D.22}$$

Putting this together with Eq. (D.16) we get

$$[A, : e^B :] =: [A, e^B] : +([A^+, B^-] - [B^+, A^-]) : e^B : \tag{D.23}$$

In the case of interest for deriving Eq. (4.80) we have

$$A^+ = (A^-)^\dagger, \quad B^+ = -(B^-)^\dagger \tag{D.24}$$

then

$$[A^+, B^-] - [B^+, A^-] = [A^+, B^-] + [A^+, B^-]^\dagger \tag{D.25}$$

With $A = \phi(y)$ and $: e^B := \psi(x)$, $[A^+, B^-]$ is purely imaginary and the right-hand side vanishes. Then Eq. (D.23) gives the identity

$$[A, : e^B :] =: [A, e^B] : \tag{D.26}$$

Appendix E

Variation of the determinant

If M is a matrix function and δM is a small variation of M, we wish to find the variation of the determinant of M (we follow Section 4.7 of [177]).

Consider

$$\delta[\ln(\mathrm{Det}M(x))] = \ln(\mathrm{Det}(M + \delta M)) - \ln(\mathrm{Det}(M))$$

$$= \ln\left[\frac{\mathrm{Det}(M + \delta M)}{\mathrm{Det}(M)}\right]$$

$$= \ln(\mathrm{Det}M^{-1}\mathrm{Det}(M + \delta M))$$

$$= \ln(\mathrm{Det}\{M^{-1}(M + \delta M)\})$$

$$= \ln(\mathrm{Det}\{1 + M^{-1}\delta M\})$$

$$= \ln(1 + \mathrm{Tr}\{M^{-1}\delta M\}) + O((\delta M)^2)$$

$$= \mathrm{Tr}\{M^{-1}\delta M\} + O((\delta M)^2) \tag{E.1}$$

Hence

$$\delta[(\mathrm{Det}M(x))] = \mathrm{Tr}[M^{-1}(x)\delta M(x)]\mathrm{Det}M(x) \tag{E.2}$$

which is the desired result.

Appendix F

Summary of cosmological equations

Assuming an isotropic and homogeneous universe, the cosmological line element is the Friedman-Robertson-Walker line element, and can be written as:

$$ds^2 = dt^2 - a^2(t)\left[\frac{dr^2}{1-kr^2} + r^2(d\theta^2 + \sin^2\theta d\phi^2)\right] \tag{F.1}$$

The function $a(t)$ is known as the scale factor. k is a parameter that is -1 for a hyperbolic or negatively curved universe, 0 for a flat universe, and $+1$ for a positively curved universe.

The equations of motion for a are derived from Einstein's equations assuming that the universe is filled with one or more fluids[1] with total energy density ρ and pressure p. Then:

$$H^2 \equiv \left(\frac{\dot{a}}{a}\right)^2 = \frac{8\pi G}{3}\rho - \frac{k}{a^2} \tag{F.2}$$

$$\ddot{a} = -\frac{4\pi G}{3}(\rho + 3p)a \tag{F.3}$$

and energy-momentum conservation gives:

$$\frac{d}{da}(\rho a^3) = -3pa^2 \tag{F.4}$$

In addition, we need the equation of state for the fluid to connect p and ρ. Some examples of equations of state are: $p = -\rho$ (cosmological constant), $p = 0$ (dust), $p = \rho/3$ (radiation), and $p = -2\rho/3$ (slowly evolving wall network). Note that ρ may contain contributions from a large variety of forms of matter and then the corresponding energy densities and pressures must be added together.

---

[1] The fluid approximation means that the relaxation time of the various components – which in our case may be plasma, gas, stars, galaxies, walls, fundamental particles – is much shorter than the characteristic time for changes in the scale factor.

Assuming a single dominant component of energy density in a flat universe ($k = 0$), Eqs. (F.2) and (F.4) can be solved to obtain:

$$p = -\rho \rightarrow a \propto e^{Ht}, \quad \rho \propto a^0$$

$$p = 0 \rightarrow a \propto t^{2/3}, \quad \rho \propto \frac{1}{a^3}$$

$$p = \frac{\rho}{3} \rightarrow a \propto t^{1/2}, \quad \rho \propto \frac{1}{a^4}$$

$$p = -\frac{2}{3}\rho \rightarrow a \propto t^2, \quad \rho \propto \frac{1}{a} \tag{F.5}$$

$$p = w\rho \rightarrow a \propto t^{2/3(w+1)}, \quad \rho \propto a^{-3(w+1)} \tag{F.6}$$

References

[1] Ablowitz, M. J. and Clarkson, P. A., *Solitons, Nonlinear Evolution Equations and Inverse Scattering*. London Mathematical Society **149** (Cambridge: Cambridge University Press, 1991).

[2] Achucarro, A. and Vachaspati, T., Semilocal and electroweak strings. *Phys. Rep.*, **327** (2000), 347. [*Phys. Rep.*, **327** (2000), 427] [arXiv:hep-ph/9904229].

[3] Actor, A., Classical solutions of SU(2) Yang-Mills theories. *Rev. Mod. Phys.*, **51** (1979), 461.

[4] Anninos, P., Oliveira, S. and Matzner, R. A., Fractal structure in the scalar lambda $(\phi^2 - 1)^2$ theory. *Phys. Rev.*, **D44** (1991), 1147.

[5] Antunes, N. D., Bettencourt, L. M. A. and Hindmarsh, M., The thermodynamics of cosmic string densities in U(1) scalar field theory. *Phys. Rev. Lett.*, **80** (1998), 908. [arXiv:hep-ph/9708215].

[6] Antunes, N. D., Pogosian, L. and Vachaspati, T., On formation of domain wall lattices. *Phys. Rev.*, **D69** (2004), 043513. [arXiv:hep-ph/0307349].

[7] Antunes, N. D. and Vachaspati, T., Spontaneous formation of domain wall lattices in two spatial dimensions. *Phys. Rev.*, **D70** (2004), 063516. [arXiv:hep-ph/0404227].

[8] Antunes, N. D., Gandra, P. and Rivers, R. J., Domain formation: decided before, or after the transition? (2005) [hep-ph/0504004].

[9] Arodz, H., Expansion in the width and collective dynamics of a domain wall. *Nucl. Phys.*, **B509** (1997), 273. [arXiv:hep-th/9703168].

[10] Arodz, H., Dziarmaga, J. and Zurek, W. H., *Patterns of Symmetry Breaking* (Dordrecht: Kluwer Academic Publishers, 2003).

[11] Astier, P., Guy, J., Regnault, N., *et al.* The supernova legacy survey: measurement of Ω_M, Ω_Λ and w from the first year data set. (2005) [astro-ph/0510447].

[12] Atiyah, M. F., Hitchin, N. J., Drinfeld, V. G. and Manin, Y. I., Construction of instantons. *Phys. Lett.*, **A65** (1978), 185.

[13] Avelino, P. P., Oliveira, J. C. R. and Martins, C. J. A., Understanding domain wall network evolution. *Phys. Lett.*, **B610** (2005), 1. [arXiv:hep-th/0503226].

[14] Avelino, P. P., Martins, C. J. A. and Oliveira, J. C. R., One-scale model for domain wall network evolution. *Phys. Rev.*, **D72** (2005), 083506. [arXiv:hep-ph/0507272].

[15] Barr, S. M. and Matheson, A. M., Weak resistance in superconducting cosmic strings. *Phys. Rev.*, **D36** (1987), 2905.

[16] Basu, R., Guth, A. H. and Vilenkin, A., Quantum creation of topological defects during inflation. *Phys. Rev.*, **D44** (1991), 340.

[17] Basu, R. and Vilenkin, A., Nucleation of thick topological defects during inflation. *Phys. Rev.*, **D46** (1992), 2345.

[18] Basu, R. and Vilenkin, A., Evolution of topological defects during inflation. *Phys. Rev.*, **D50** (1994), 7150. [arXiv:gr-qc/9402040].

[19] Bogolubsky, I. L. and Makhanov, V. G., Lifetime of pulsating solitons in certain classical models. *JETP Lett.*, **24** (1976), 12.

[20] Bogomolnyi, E. B., The stability of classical solutions. *Sov. J. Nucl. Phys.*, **24** (1976), 449.

[21] Bonjour, F., Charmousis, C. and Gregory, R., The dynamics of curved gravitating walls. *Phys. Rev.*, **D62** (2000), 083504. [arXiv:gr-qc/0002063].

[22] Borde, A., Trodden, M. and Vachaspati, T., Creation and structure of baby universes in monopole collisions. *Phys. Rev.*, **D59** (1999), 043513. [arXiv:gr-qc/9808069].

[23] Boys, C. V., *Soap Bubbles: Their Colors and Forces Which Mold Them* (New York: Dover Publications, 1959).

[24] Bray, A., Topological defects and phase ordering dynamics. In *Formation and Interaction of Topological Defects*, ed. A.-C. Davis and R. Brandenberger (New York: NATO ASI Series, Plenum, 1995).

[25] Bucher, M. and Spergel, D. N., Is the dark matter a solid? *Phys. Rev.*, **D60** (1999), 043505. [arXiv:astro-ph/9812022].

[26] Campbell, D. K., Schonfeld, J. F. and Wingate, C. A., Resonance structure in kink-antikink interactions in phi**4 theory. *Physica*, **9D** (1983), 1.

[27] Caroli, C., de Gennes, P. G. and Matricon, J., Bound fermion states on a vortex line in a type II superconductor. *Phys. Lett.*, **9** (1964), 307.

[28] Carter, B. and Gregory, R., Curvature corrections to dynamics of domain walls. *Phys. Rev.*, **D51** (1995), 5839. [arXiv:hep-th/9410095].

[29] Chamblin, A. and Eardley, D. M., Puncture of gravitating domain walls. *Phys. Lett.*, **B475** (2000), 46. [arXiv:hep-th/9912166].

[30] Cheng, T. P. and Li, L. F., *Gauge Theory of Elementary Particle Physics* (New York: Oxford University Press, 1984).

[31] Cho, I. and Vilenkin, A., Spacetime structure of an inflating global monopole. *Phys. Rev.*, **D56** (1997), 7621. [arXiv:gr-qc/9708005].

[32] Ciria, J. C. and Tarancon, A., Renormalization group study of the soliton mass in the $(1+1)$-dimensional $\lambda\phi^4$ lattice model. *Phys. Rev.*, **D49** (1994), 1020. [arXiv:hep-lat/9309019].

[33] Coleman, S. R. and Weinberg, E., Radiative corrections as the origin of spontaneous symmetry breaking. *Phys. Rev.*, **D7** (1973), 1888.

[34] Coleman, S. R., Quantum sine-Gordon equation as the massive Thirring model. *Phys. Rev.*, **D11** (1975), 2088.

[35] Coleman, S., *Aspects of Symmetry* (Cambridge: Cambridge University Press, 1985).

[36] Coulson, D., Lalak, Z. and Ovrut, B. A., Biased domain walls. *Phys. Rev.*, **D53** (1996), 4237.

[37] Creutz, M., *Quarks, Gluons and Lattices* (Cambridge: Cambridge University Press, 1985).

[38] Dashen, R. F., Hasslacher, B. and Neveu, A., Nonperturbative methods and extended hadron models in field theory. 1. Semiclassical functional methods. *Phys. Rev.*, **D10** (1974), 4114.

[39] Dashen, R. F., Hasslacher, B. and Neveu, A., Nonperturbative methods and extended hadron models in field theory. 2. Two-dimensional models and extended hadrons. *Phys. Rev.*, **D10** (1974), 4130.

[40] Dashen, R. F., Hasslacher, B. and Neveu, A., Nonperturbative methods and extended hadron models in field theory. 3. Four-dimensional nonabelian models. *Phys. Rev.*, **D10** (1974), 4138.

[41] Dashen, R. F., Hasslacher, B. and Neveu, A., The particle spectrum in model field theories from semiclassical functional integral techniques. *Phys. Rev.*, **D11** (1975), 3424.

[42] Dashen, R. F., Hasslacher, B. and Neveu, A., Semiclassical bound states in an asymptotically free theory. *Phys. Rev.*, **D12** (1975), 2443.

[43] Davidson, A., Toner, B. F., Volkas, R. R. and Wali, K. C., Clash of symmetries on the Brane. *Phys. Rev.*, **D65** (2002), 125013. [arXiv:hep-th/0202042].

[44] von Delft, J. and Schoeller, H., Bosonization for beginners: Refermionization for experts. *Ann. Phys.*, **7** (1998), 225. [arXiv:cond-mat/9805275].

[45] Derrick, G. H., Comments on nonlinear wave equations as models for elementary particles. *J. Math. Phys.*, **5** (1964), 1252.

[46] Deser, S., Plane waves do not polarize the vacuum. *J. Phys. A, Math. Gen.*, **8** (1975), 1972.

[47] Dolan, L. and Jackiw, R., Symmetry behavior at finite temperature. *Phys. Rev.*, **D9** (1974), 3320.

[48] Drazin, P. and Johnson, R., *Solitons: An Introduction* (Cambridge: Cambridge University Press, 1989).

[49] Drell, S. D., Weinstein, M. and Yankielowicz, S., Variational approach to strong coupling field theory. 1. Φ^4 theory. *Phys. Rev.*, **D14** (1976), 487.

[50] Dvali, G. R., Melfo, A. and Senjanovic, G., Is there a monopole problem? *Phys. Rev. Lett.*, **75** (1995), 4559. [arXiv:hep-ph/9507230].

[51] Dvali, G. R. and Shifman, M. A., Domain walls in strongly coupled theories. *Phys. Lett.*, **B396** (1997), 64. [Erratum-*ibid.* **B407** (1997), 452] [arXiv:hep-th/9612128].

[52] Dvali, G. R., Liu, H. and Vachaspati, T., Sweeping away the monopole problem. *Phys. Rev. Lett.*, **80** (1998), 2281. [arXiv:hep-ph/9710301].

[53] Everett, A. E., Observational consequences of a "domain" structure of the universe. *Phys. Rev.*, **D 10** (1974), 3161.

[54] Feynman, R. P., Leighton, R. B. and Sands, M., *The Feynman Lectures on Physics, Volume II* (Reading, MA: Addison-Wesley Publishing Company, 1964).

[55] Feynman, R. P., Leighton, R. B. and Sands, M., *The Feynman Lectures on Physics, Volume III* (Reading, MA: Addison-Wesley Publishing Company, 1965).

[56] Fordy, A. P., ed., *Soliton Theory: A Survey of Results* (Manchester: Manchester University Press, 1990).

[57] Forster, D., Dynamics of relativistic vortex lines and their relation to dual theory. *Nucl. Phys.*, **B81** (1974), 84.

[58] Friedlander, F. G., *The Wave Equation on a Curved Space-time* (Cambridge: Cambridge University Press, 1976).

[59] Garagounis, T. and Hindmarsh, M., Scaling in numerical simulations of domain walls. *Phys. Rev.*, **D68** (2003), 103506. [arXiv:hep-ph/0212359].

[60] Gelmini, G. B., Gleiser, M. and Kolb, E. W., Cosmology of biased discrete symmetry breaking. *Phys. Rev.*, **D39** (1989), 1558.

[61] de Gennes, P. G., *Superconductivity of Metals and Alloys* (Oxford: Perseus Books Publishing, 1999).

[62] Georgi, H., *Lie Algebras in Particle Physics* (Reading, MA: Benjamin/Cummings Publishing Company, 1982).

[63] Georgi, H. and Glashow, S. L., Unity of all elementary particle forces. *Phys. Rev. Lett.*, **32** (1974), 438.

[64] Gleiser, M., Pseudostable bubbles. *Phys. Rev.*, **D49** (1994), 2978. [arXiv:hep-ph/9308279].

[65] Goldenfeld, N., Dynamics of cosmological phase transitions: what can we learn from condensed matter physics? In *Formation and Interaction of Topological Defects*, ed. A.-C. Davis and R. Brandenberger (New York: NATO ASI Series, Plenum, 1995).

[66] Goldhaber, A. S., Rebhan, A., van Nieuwenhuizen, P. and Wimmer, R., Quantum corrections to mass and central charge of supersymmetric solitons. *Phys. Rep.*, **398** (2004), 179. [arXiv:hep-th/0401152].

[67] Goldstone, J. and Jackiw, R., Quantization of nonlinear waves. *Phys. Rev.*, **D11** (1975), 1486.

[68] Goldstone, J. and Wilczek, F., Fractional quantum numbers on solitons. *Phys. Rev. Lett.*, **47** (1981), 986.

[69] Gomez Nicola, A. and Steer, D. A., Thermal bosonisation in the sine-Gordon and massive Thirring models. *Nucl. Phys.*, **B549** (1999), 409. [arXiv:hep-ph/9810519].

[70] Gózdz, W. T. and Holyst, R., Triply periodic surfaces and multiply continuous structures from the Landau model of microemulsions. *Phys. Rev.*, **E54** (1996), 5012.

[71] Gradshteyn, I. S. and Ryzhik, M., *Table of Integrals, Series, and Products* (New York: Academic Press, 1980).

[72] Gregory, R., Effective action for a cosmic string. *Phys. Lett.*, **B206** (1988), 199.

[73] Gregory, R., Haws, D. and Garfinkle, D., The dynamics of domain walls and strings. *Phys. Rev.*, **D42** (1990), 343.

[74] Harvey, J. A., Kolb, E. W., Reiss, D. B. and Wolfram, S., Calculation of cosmological baryon asymmetry in grand unified gauge models. *Nucl. Phys.*, **B201** (1982), 16.

[75] Heeger, A. J., Kivelson, S., Schrieffer, J. R. and Su, W.-P., Solitons in conducting polymers. *Rev. Mod. Phys.*, **60** (1988), 781.

[76] Hill, C. T., Schramm, D. N. and Fry, J. N., Cosmological structure formation from soft topological defects. *Comments Nucl. Part. Phys.*, **19** (1989), 25.

[77] Hindmarsh, M., Analytic scaling solutions for cosmic domain walls. *Phys. Rev. Lett.*, **77** (1996), 4495. [arXiv:hep-ph/9605332].

[78] Hindmarsh, M., Evolution of defect and brane networks. *Phys. Rev.*, **D68** (2003), 043510. [arXiv:hep-ph/0207267].

[79] 't Hooft, G., Magnetic monopoles in unified gauge theories. *Nucl. Phys.*, **B79** (1974), 276.

[80] Ipser, J. and Sikivie, P., The gravitationally repulsive domain wall. *Phys. Rev.*, **D30** (1984), 712.

[81] Israel, W., Singular hypersurfaces and thin shells in general relativity. *Nuovo Cimento B*, **44S10** (1966), 1. [Erratum-*ibid.* **48** (1967), 463.]

[82] Iwazaki, A., Ferromagnetic domain wall and primeval magnetic field. *Phys. Rev.*, **D56** (1997), 2435. [arXiv:hep-ph/9608448].

[83] Jackiw, R. and Rebbi, C., Solitons with fermion number 1/2. *Phys. Rev.*, **D13** (1976), 3398.

[84] Jackiw, R. and Rossi, P., Zero modes of the vortex-fermion system. *Nucl. Phys.*, **B190** (1981), 681.

[85] Jona, F. and Shirane, G., *Ferroelectric Crystals* (Oxford: Pergamon, 1962).

[86] Karra, G. and Rivers, R. J., Initial vortex densities after a temperature quench. *Phys. Lett.*, **B414** (1997), 28. [arXiv:hep-ph/9705243].

[87] Kibble, T. W. B., Topology of cosmic domains and strings. *J. Phys.*, **A9** (1976), 1387.

[88] Kibble, T. W. B., Lazarides, G. and Shafi, Q., Walls bounded by strings. *Phys. Rev.*, **D26** (1982), 435.

[89] Kirzhnits, D. A. and Linde, A. D., A relativistic phase transition. *Sov. Phys.-JETP*, **40** (1974), 628.

[90] Kirzhnits, D. A. and Linde, A. D., Symmetry behavior in gauge theories. *Ann. Phys.*, **101** (1976), 195.

[91] Kosowsky, A., Turner, M. S. and Watkins, R., Gravitational radiation from colliding vacuum bubbles. *Phys. Rev.*, **D45** (1992), 4514.

[92] Kubotani, H., The domain wall network of explicitly broken O(N) model. *Prog. Theor. Phys.*, **87** (1992), 387.

[93] Laguna, P. and Zurek, W. H., Density of kinks after a quench: When symmetry breaks, how big are the pieces? *Phys. Rev. Lett.*, **78** (1997), 2519. [arXiv:gr-qc/9607041].

[94] de Laix, A. A., Trodden, M. and Vachaspati, T., Topological inflation with multiple winding. *Phys. Rev.*, **D57** (1998), 7186. [arXiv:gr-qc/9801016].

[95] de Laix, A. A. and Vachaspati, T., On random bubble lattices. *Phys. Rev.*, **D59** (1999), 045017. [arXiv:hep-ph/9802423].

[96] Lalak, Z. and Ovrut, B. A., Domain walls, percolation theory and Abell clusters. *Phys. Rev. Lett.*, **71** (1993), 951.

[97] Larsson, S. E., Sarkar, S. and White, P. L., Evading the cosmological domain wall problem. *Phys. Rev.*, **D55** (1997), 5129. [arXiv:hep-ph/9608319].

[98] Lazarides, G. and Shafi, Q., Superconducting membranes. *Phys. Lett.*, **B159** (1985), 261.

[99] Li, L. F., Group theory of the spontaneously broken gauge symmetries. *Phys. Rev.*, **D9** (1974), 1723.

[100] Linde, A. D., Phase transitions in gauge theories and cosmology. *Rep. Prog. Phys.*, **42** (1979), 389.

[101] Linde, A. D., Monopoles as big as a universe. *Phys. Lett.*, **B327** (1994), 208. [arXiv:astro-ph/9402031].

[102] Linde, A. D. and Linde, D. A., Topological defects as seeds for eternal inflation. *Phys. Rev.*, **D50** (1994), 2456. [arXiv:hep-th/9402115].

[103] Liu, H. and Vachaspati, T., SU(5) monopoles and the dual standard model. *Phys. Rev.*, **D56** (1997), 1300. [arXiv:hep-th/9604138].

[104] Lowe, M. J. and Osborn, H., Bound for soliton masses in two dimensional field theories. Unpublished KEK library preprint (1977).

[105] Maeda, K. I. and Turok, N., Finite width corrections to the Nambu action for the Nielsen-Olesen string. *Phys. Lett.*, **B202** (1988), 376.

[106] Malomed, B. A., Emission from, WKB quantization, and stochastic decay of sine-Gordon solitons in external fields. *Phys. Lett.*, **A120** (1987), 28.

[107] Malomed, B. A., Decay of shrinking solitons in multidimensional sine-Gordon equation. *Physica*, **24D** (1987), 155.

[108] Mandelstam, S., Soliton operators for the quantized sine-Gordon equation. *Phys. Rev.*, **D11** (1975), 3026.

[109] Manton, N. S., An effective Lagrangian for solitons. *Nucl. Phys.*, **B150** (1979), 397.

[110] Manton, N. S. and Merabet, H., ϕ^4 kinks - gradient flow and dynamics. *Nonlinearity*, **12** (1996), 851. [arXiv:hep-th/9605038].

[111] Mazenko, G. F., Introduction to growth kinetic problems. In *Formation and Interaction of Topological Defects*, ed. A.-C. Davis and R. Brandenberger (New York: NATO ASI Series, Plenum, 1995).

[112] Montvay, I. and Münster, G., *Quantum Fields on a Lattice* (Cambridge: Cambridge University Press, 1994).

[113] Morse, P. M. and Feshbach, H., *Methods of Theoretical Physics* (New York: McGraw-Hill, 1953).

[114] Nahm, W., A simple formalism for the bps monopole. *Phys. Lett.*, **B90** (1980), 413.

[115] Nitsche, J. C. C., *Lectures on Minimal Surfaces, Vol. I* (Cambridge: Cambridge University Press, 1989).

[116] Ohta, T., Jasnow, D. and Kawasaki, K., Universal scaling in the motion of random interfaces. *Phys. Rev. Lett.*, **49** (1982), 1223.

[117] Oprea, J., *The Mathematics of Soap Films: Explorations with Maple* (Providence, RI: American Mathematical Society, 2000).

[118] Perlmutter, S., Aldering, G., Goldhaber, G., *et al.*, Measurements of omega and lambda from 42 high-redshift supernovae. *Astrophys. J.*, **517** (1999), 565. [arXiv:astro-ph/9812133].

[119] Peskin, M. E. and Schroeder, D. V., *Quantum Field Theory* (Boulder, CO: Westview Press, 1995).

[120] Pogosian, L. and Vachaspati, T., Space of kink solutions in SU(N) x Z(2). *Phys. Rev.*, **D64** (2001), 105023. [arXiv:hep-th/0105128].

[121] Pogosian, L., Kink interactions in SU(N) x Z(2). *Phys. Rev.*, **D65** (2002), 065023. [arXiv:hep-th/0111206].

[122] Pogosian, L., Steer, D. A. and Vachaspati, T., Triplication of SU(5) monopoles. *Phys. Rev. Lett.*, **90** (2003), 061801. [arXiv:hep-th/0204106].

[123] Pogosian, L. and Vachaspati, T., Domain wall lattices. *Phys. Rev.*, **D67** (2003), 065012. [arXiv:hep-th/0210232].

[124] Polyakov, A. M., Particle spectrum in quantum field theory. *JETP Lett.*, **20** (1974), 194.

[125] Press, W. H., Ryden, B. S. and Spergel, D. N., Dynamical evolution of domain walls in an expanding universe. *Astrophys. J.*, **347** (1989), 590.

[126] Rajaraman, R., *Solitons and Instantons* (Amsterdam: North-Holland, 1982).

[127] Riess, A. G., Filippenks, A. V., Challis, P., *et al.*, Observational evidence from supernovae for an accelerating universe and a cosmological constant. *Astron. J.*, **116** (1998), 1009. [arXiv:astro-ph/9805201].

[128] Riess, A. G., Strolger, L.-G., Tonny, J., *et al.*, Type Ia supernova discoveries at z > 1 from the Hubble Space Telescope: Evidence for past deceleration and constraints on dark energy evolution. *Astrophys. J.*, **607** (2004), 665. [arXiv:astro-ph/0402512].

[129] Rivier, N., Order and disorder in packings and froths. In *Disorder and Granular Media*, ed. D. Bideau and A. Hansen (Amsterdam: North-Holland, 1993).

[130] Ryden, B. S., Press, W. H. and Spergel, D. N., The evolution of networks of domain walls and cosmic strings. *Astrophys. J.*, **357** (1990), 293.

[131] Sakai, N., Shinkai, H. A., Tachizawa, T. and Maeda, K. I., Dynamics of topological defects and inflation. *Phys. Rev.*, **D53** (1996), 655. [Erratum-*ibid.* **D54** (1996), 2981.] [arXiv:gr-qc/9506068].

[132] Salle, M., Kinks in the Hartree approximation. *Phys. Rev.*, **D69** (2004), 025005. [arXiv:hep-ph/0307080].

[133] Scalapino, D. J. and Stoeckly, B., Quantum mechanical kink solution. *Phys. Rev.*, **D14** (1976), 3376.

[134] Scott, A. C., A nonlinear Klein-Gordon equation. *Am. J. Phys.*, **37** (1969), 52.

[135] Segur, H. and Kruskal, M. D., Nonexistence of small amplitude breather solutions in Φ^4 theory. *Phys. Rev. Lett.*, **58** (1987), 747.

[136] Shellard, E. P. S., *Quantum Effects in the Early Universe*. Unpublished Ph.D. thesis, University of Cambridge (1986).

[137] Sheng, Q. and Elser, V., Quasicrystalline minimal surfaces. *Phys. Rev.*, **B49** (1994), 9977.

[138] Silveira, V., Dynamics of the lambda Φ^4 kink. *Phys. Rev.*, **D38** (1988), 3823.

[139] Simon, B., The bound state of weakly coupled Schrödinger operators in one and two-dimensions. *Ann. Phys.*, **97** (1976), 279.

[140] Slusarczyk, M., Dynamics of a planar domain wall with oscillating thickness in $\lambda\phi^4$ model. *Acta Phys. Polon.*, **B31** (2000), 617. [arXiv:hep-th/9903185].

[141] Smit, J., *Introduction to Quantum Fields on a Lattice* (Cambridge: Cambridge University Press, 2002).

[142] Son, D. T., Stephanov, M. A. and Zhitnitsky, A. R., Domain walls of high-density QCD. *Phys. Rev. Lett.*, **86** (2001), 3955. [arXiv:hep-ph/0012041].

[143] Stauffer, D., Scaling theory of percolation clusters. *Phys. Rep.*, **54** (1979), 1.

[144] Stebbins, A. and Turner, M. S., Is the great attractor really a great wall? *Astrophys. J.*, **339** (1989), L13.

[145] Steenrod, N., *The Topology of Fiber Bundles* (Princeton: Princeton University Press, 1951).

[146] Steer, D. and Vachaspati, T., Domain walls and fermion scattering in grand unified models. Unpublished (2005).

[147] Stojkovic, D., Fermionic zero modes on domain walls. *Phys. Rev.*, **D63** (2001), 025010. [arXiv:hep-ph/0007343].

[148] Stojkovic, D., Freese, K. and Starkman, G. D., Holes in the walls: Primordial black holes as a solution to the cosmological domain wall problem. *Phys. Rev.*, **D72** (2005), 045012. [arXiv:hep-ph/0505026].

[149] Su, W. P., Schrieffer, J. R. and Heeger, A. J., Solitons in polyacetylene. *Phys. Rev. Lett.*, **42** (1979), 1698.

[150] Su, W. P., Schrieffer, J. R. and Heeger, A. J., Soliton excitations in polyacetylene. *Phys. Rev.*, **B22** (1980), 2099.

[151] Su, W. P. and Schrieffer, J. R., Fractionally charged excitations in charge-density-wave systems with commensurability 3. *Phys. Rev. Lett.*, **46** (1980), 738.

[152] Su, W. P. and Schrieffer, J. R., Soliton dynamics in polyacetylene. *Proc. Natl. Acad. Sci. USA*, **77** (1980), 5626.

[153] Sutcliffe, P. M., Classical and quantum kink scattering. *Nucl. Phys.*, **B393** (1993), 211.

[154] Taub, A. H., Isentropic hydrodynamics in plane symmetric spacetimes. *Phys. Rev.*, **103** (1956), 454.

[155] Toda, M., *Theory of Nonlinear Lattices* (New York: Springer-Verlag, 1989).

[156] Torres-Hernandez, J. L., The soliton mass in the sine-Gordon theory. *Nucl. Phys.*, **B116** (1976), 214.

[157] Turner, M. S. and Wilczek, F., Relic gravitational waves and extended inflation. *Phys. Rev. Lett.*, **65** (1990), 3080.

[158] Turner, M. S., Watkins, R. and Widrow, L. M., Microwave distortions from collapsing domain wall bubbles. *Astrophys. J.*, **367** (1991), L43.

[159] Vachaspati, T. and Vilenkin, A., Formation and evolution of cosmic strings. *Phys. Rev.*, **D30** (1984), 2036.

[160] Vachaspati, T., Everett, A. E. and Vilenkin, A., Radiation from vacuum strings and domain walls. *Phys. Rev.*, **D30** (1984), 2046.

[161] Vachaspati and Vachaspati, T., Traveling waves on domain walls and cosmic strings. *Phys. Lett.*, **B238** (1990), 41.

[162] Vachaspati, T., An attempt to construct the standard model with monopoles. *Phys. Rev. Lett.*, **76** (1996), 188. [arXiv:hep-ph/9509271].

[163] Vachaspati, T., A class of kinks in SU(N) x Z(2). *Phys. Rev.*, **D63** (2001), 105010. [arXiv:hep-th/0102047].

[164] Vachaspati, T., Symmetries within domain walls. *Phys. Rev.*, **D67** (2003), 125002. [arXiv:hep-th/0303137].

[165] Vachaspati, T., Reconstruction of field theory from excitation spectra of defects. *Phys. Rev.*, **D69** (2004), 043510. [arXiv:hep-th/0309086].

[166] Vachaspati, T. and Trodden, M., Causality and cosmic inflation. *Phys. Rev.*, **D61** (2000), 023502. [arXiv:gr-qc/9811037].

[167] Vilenkin, A., Gravitational field of vacuum domain walls and strings. *Phys. Rev.*, **D23** (1981), 852.

[168] Vilenkin, A. and Everett, A. E., Cosmic strings and domain walls in models with Goldstone and pseudogoldstone bosons. *Phys. Rev. Lett.*, **48** (1982), 1867.

[169] Vilenkin, A., Gravitational field of vacuum domain walls. *Phys. Lett.*, **B133** (1983), 177.

[170] Vilenkin, A., Topological inflation. *Phys. Rev. Lett.*, **72** (1994), 3137. [arXiv:hep-th/9402085].

[171] Vilenkin, A. and Shellard, E. P. S., *Cosmic Strings and other Topological Defects* (Cambridge: Cambridge University Press, 1994).

[172] Voloshin, M. B., No primordial magnetic field from domain walls. *Phys. Lett.*, **B491** (2000), 311. [arXiv:hep-ph/0007123].

[173] Voloshin, M. B., Once again on electromagnetic properties of a domain wall interacting with charged fermions. *Phys. Rev.*, **D63** (2001), 125012. [arXiv:hep-ph/0102239].

[174] Volovik, G. E., *The Universe in a Helium Droplet* (Oxford: Oxford University Press, 2003).

[175] Watkins, R. and Widrow, L. M., Aspects of reheating in first order inflation. *Nucl. Phys.*, **B374** (1992), 446.

[176] Weinberg, E. J., Index calculations for the fermion - vortex system. *Phys. Rev.*, **D24** (1981), 2669.

[177] Weinberg, S., *Gravitation and Cosmology* (New York: John Wiley & Sons, 1972).

[178] Weinberg, S., Perturbative calculations of symmetry breaking. *Phys. Rev.*, **D7** (1973), 2887.

[179] Weinberg, S., Gauge and global symmetries at high temperature. *Phys. Rev.*, **D9** (1974), 3357.

[180] Whitham, G. B., *Linear and Nonlinear Waves* (New York: John Wiley & Sons, 1974).

[181] Widrow, L. M., General relativistic domain walls. *Phys. Rev.*, **D39** (1989), 3571.

[182] Widrow, L. M., The collapse of nearly spherical domain walls. *Phys. Rev.*, **D39** (1989), 3576.

[183] Widrow, L. M., Dynamics of thick domain walls. *Phys. Rev.*, **D40** (1989), 1002.

[184] Witten, E., Superconducting strings. *Nucl. Phys.*, **B249** (1984), 557.

[185] Zabusky, N. J. and Kruskal, M. D., Interactions of "solitons" in a collisionless plasma and the recurrence of initial states. *Phys. Rev. Lett.*, **15** (1965), 240.

[186] Zeldovich, Y. B., Kobzarev, I. Y. and Okun, L. B., Cosmological consequences of the spontaneous breakdown of discrete symmetry. *Zh. Eksp. Teor. Fiz.*, **67** (1974), 3.

[187] Zurek, W. H., Cosmic strings in laboratory superfluids and the topological remnants of other phase transitions. *Acta Phys. Polon.*, **B24** (1993), 1301.

[188] Zurek, W. H., Cosmological experiments in condensed matter systems. *Phys. Rep.*, **276** (1996), 177. [arXiv:cond-mat/9607135].

Index